50 Model Rocket Projects for the Evil Genius

Evil Genius Series

50 Model Rocket Projects for the Evil Genius

GAVIN D. J. HARPER

McGraw-Hill

New York   Chicago   San Francisco   Lisbon
London   Madrid   Mexico City   Milan   New Delhi
San Juan   Seoul   Singapore   Sydney   Toronto

The McGraw·Hill Companies

Library of Congress Cataloging-in-Publication Data

Harper, Gavin D. J.
    50 model rocket projects for the evil genius / Gavin D. J. Harper.
        p.   cm. (Evil genius series)
    Includes index.
    ISBN 0-07-146984-2 (alk. paper)
    1. Rockets (Aeronautics)—Models—Juvenile literature. I. Title. II. Title:
Fifty model rocket projects for the evil genius. III. Series.

TL844.H37   2006
621.43'560228—dc22   2006015600

1 2 3 4 5 6 7 8 9 0   QPD/QPD   0 1 2 1 0 9 8 7 6

ISBN-13: 978-0-07-146984-5
ISBN-10: 0-07-146984-2

The sponsoring editor for this book was Judy Bass and the production supervisor was Pamela A. Pelton. It was set in Times Ten by Keyword Group Ltd. The art director for the cover was Anthony Landi.

Printed and bound by Quebecor/Dubuque.

McGraw-Hill books are available at special quantity discounts to use as premiums and sales promotions, or for use in corporate training programs. For more information, please write to the Director of Special Sales, McGraw-Hill Professional, Two Penn Plaza, New York, NY 10121-2298. Or contact your local bookstore.

This book is printed on acid-free paper.

Contents

For Rodney Buckland

Who helped me reach for the stars

Gavin D. J. Harper is author of *50 Awesome Auto Projects for the Evil Genius, Build Your Own Car PC*, and the forthcoming *Solar Energy Projects for the Evil Genius* (all from McGraw-Hill), and has had work mentioned in the journal *Science*. He is a Science and Engineering Ambassador for the SETNET group and is passionate about engaging young people in the science, engineering, and technical hobbies that have given him so much fun. He has been building model rockets, specialized computers, and other scientific projects for years. Mr. Harper is currently completing his BSc. (Hons.) Technology with the Open University and his MSc. Architecture: Advanced Environmental & Energy Studies with the University of East London and the Centre for Alternative Technology. He lives in Essex, England.

Acknowledgments

First, a big thanks to the legendary Bob Spary, my diesel Mercedes driving über-Physics teacher, who tried to put me off investigating rocket parachutes for the GCSE Science exam because the subject only had tenuous links with the marking scheme. Despite that fact, I went ahead, and my marks were grim; thanks for putting up with my going off at tangents, and trying to find some way of giving me credit for them.

Sincerest thanks to Peter Barratt and the rest of the HART team who encouraged me when I was starting out in model rocketry. Peter now sells photos of model rocket launches to raise money for Asthma UK. I suggest you Google for him and support the cause.

Thanks also go to Andrew Stainer for his help in taking some of the photos in this book. I think we both agreed that it was a fantastic day at the park scaring the kids.

A special thank you goes to the Wade family from Upminster, UK, who were kind enough to pose for the pictures in the Introduction while walking in the park on their son's birthday. The shock and awe reactions are genuine ☺.

My thanks to Dr. Ray Wallace and Dr. Stuart Bennett, who introduced me to the making of hydrogen pipette rockets at the Open University ST240 Summer School, also to my partner Mike at the Nottingham Summer School, who tolerated my limited accuracy in weighing chemicals. (I chose to err on the side of caution with "more" being better than less).

Thanks also to Luke Griffiths and Mickey Sanchez from the Starchaser group, who presented a fantastic day of space activities to the kids at Hall Mead School! Your perspectives on delivering model rocket lessons in school were invaluable. Thanks also to Dino Shaheed for letting me get involved, and to the pupils at Hall Mead School for the fantastic day launching rockets and for all of your enthusiasm.

Many thanks to Gary White, President of the Out of This World Cambridge Rocket Club for the photos of club events and advice.

Paul Lavin from Hesperis Technology is to be applauded for managing to get me all of the rocket supplies for this book when they were needed yesterday, battling against the odds despite the back-end of his web shop crashing, and having to run to the Post Office with the parcel in hand!

Thanks to Tim Van Milligan (a.k.a. "Mr. Rocket") for poking me in the right direction with RockSim, and for allowing me to use screenshots from his fabulous software.

I also owe my eternal gratitude to James Padfield, who took my ideas and helped work them into a reality for Chapter 12.

On the other side of the pond thanks to Ken Gracey from Parallax Inc., who made the money-off coupons in the back of this book come to fruition. Think of Ken when you are saving cash on your BASIC Stamps.

Thanks to Mark B. Bundick, President of the National Association of Rocketry (NAR) for allowing me to reproduce the NAR Safety Code in the back of this book.

I owe Art Applewhite a great debt of gratitude for his advice on flying saucers and his permission to reproduce the graphics for the UFOs in this book.

I could not write an acknowledgements page without thanking the people behind the scenes who turn my disarray of text and images into something presentable – Andy Baxter and Non Oastler at Keyword.

Finally, my amazing editor Judy Bass, who has been thoroughly supportive throughout the process, and goes from strength to strength – always on tap to supply positivity and words of encouragement at the time they are required. Thanks Judy for continuing to be a star.

Acknowledgments

Introduction

Rockets are fascinating to kids and adults alike, and as such the prospect of ACTUALLY BEING ABLE TO LAUNCH A ROCKET IN THE YARD should fill even the biggest kids with a great amount of enthusiasm.

The first time that "rockets" captured my imagination was the first time I saw the TV television series *Thunderbirds*. My dad saw *Thunderbirds* as a kid before me when it first aired in 1965, and I can remember him sitting me down in the living room and saying "Watch this, it's great." Within about 5 minutes I was hooked, and every Wednesday night I would religiously tune in to BBC2 and watch as the excitement unfolded. It was just something about the plumes of smoke as the rockets ascended into the air, the flames licking out from the exhaust, that captured my imagination. It was only much later that I was to realise that many of the special effects were conceived by Derek Meddings using what were essentially "thrustless" model rocket motors.

Some years later, I got an Estes Alpha 3 kit for my birthday and that was it – I was hooked. From then I migrated to slightly larger and more complex kits – the great thing about this hobby is that it grows with you as your interest and your hobby develops.

It was shortly afterwards that I met Peter Barratt of the Hornchurch Airfield Rocket Team who I thank for his encouragement when I was just starting out in the hobby. Rocketry can be a very social activity, a great way of meeting new friends.

Model rocketry is a great tool to teach many topics to kids. What is even better is that as well as being a tool, model rocketry is an *incentive* to learn science; topics such as math, physics, industrial design, and ICT (Information Communication Technology), can all be taught using model rocketry as a "launch vehicle."

I learnt a lot of science from model rockets – the principles that we rely upon for their successful operation are the basic tenets of chemistry and physics – but there is plenty of room to "tailor" the hobby to your own interests. After launching several

Figure I-1 *Thunderbird 1.*

rockets I decided that the lessons I had learnt from another hobby of mine, electronics, could be easily applied to model rocketry. The fusion of these two interests results in some really exciting experiments and interesting projects.

Whether you want to look at the hobby as a serious educational tool, or as a passing folly, one thing is certain – you will learn a lot along the way.

The first part of this book is more or less wholly devoted to rocketry. It contains all of the fundamental information that you really need to know to get involved with the hobby. Here is a good place to start if you know lots about electronics but little about rocketry.

The second half of this book is more or less wholly devoted to electronics. If you are already involved with rocketry but know little about electronics, then here is a good place to start. This book assumes some familiarity with the BASIC Stamp and its programming language; if you are not already familiar with the BASIC Stamp, I can highly recommend *Programming and Customizing the BASIC Stamp Computer* by Scott Edwards. Once you get into the programming language it really is very easy, you will quickly find yourself being able to tailor the software to the data you want to be able to retrieve and collect.

Hopefully this book provides a "payload" of great ideas for you to pick up and run with; do not let the projects in this book hamper your creativity by sticking rigidly to a formula. Take the ideas, evolve them and turn them into a creation that is truly your own.

Rocketry is a great activity that can engage all members of the family – the fun and the thrill you experience by launching your own model rocket is something for everyone to share, from the anticipation of waiting for launch, to the excitement and wonder as the rocket leaves the pad.

I truly hope that you will have as much fun with the hobby as I have done, and enjoy constructing some of the projects in this book.

To paraphrase Kennedy's Moon Speech (Rice Stadium, September 12, 1962).

We choose to amuse ourselves with Model Rocketry. We choose to amuse ourselves with Model Rocketry in this decade and do the other things, not because they are easy, but because they are hard, because that goal will serve to organize and measure the best of our energies and skills, because that challenge is one that we are willing to accept, one we are unwilling to postpone, and one which we intend to win, and the others, too.

Figure I-2 *The anticipation before launch.*

Figure I-3 *Excitement and wonder as the rocket leaves the pad.*

Introduction

2

Chapter 1

History of Rocketry

The Space Age kicked off in the late 1950s; however, the history of rocketry itself is much older, going back millennia. Rocketry has a checkered past – different countries all claim that they were responsible for invention and innovation; however, what is presented here is my own "canned history" of launching things into space.

The invention of black powder, the first explosive, is hotly disputed, with the Arabs, the English alchemist Roger Bacon, the German–Greek alchemist Berthold Schwarz, the Hindus, and the Chinese all laying claim to its invention. It is not beyond the realms of possibility that black powder was invented independently and simultaneously in different countries; however, commonly, it is accepted that the Chinese started the ball rolling by inventing black powder (火藥) back in the 9th century. By the 11th century, they were busy launching fire arrows, rockets from catapults, and various forms of projectiles using black powder. Rocketry was now in its infancy. You can learn a little more about black powder in Project 1: Making Black Powder.

The first "book" on rocketry was published by the Lithuanian–Polish Kazimierz Siemienowicz and was entitled *Artis Magnae Artilleriae pars prima* which translates as *The Complete Art of Artillery*. It certainly makes for a heavier read than anything in the Evil Genius series! The book contained chapters on construction and even subjects as advanced as multi-stage rocketry. At the time, it was certainly "ahead of the curve." Figure 1-1 shows one of Siemienowicz's illustrations from the book.

This man was an absolute visionary! One totally amazing thing is the uncanny resemblance between the rocket pictured in the book, and modern-day multi-stage rockets. There is clearly a nozzle cut into the fuel, not unlike a cross-section through a model rocket motor, and it can be seen how one rocket ignites the stage above. AMAZING!

As far back as 1696, Robert Anderson published a paper *The Making of Rockets* describing how to make molds and propellants for rockets as well as how to perform some rocket calculations (see Figure 1-2).

At the Battle of Baltimore in 1814, rockets were used by HMS Erebus against Fort McHenry; the words "the rockets' red glare" were later immortalized in *The Star-Spangled Banner.*

In 1898, H.G. Wells released his seminal work *War of the Worlds;* this was the start of a public fascination

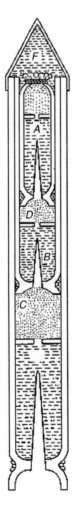

Figure 1-1 *Siemienowicz multi-stage rocket.*
Image in the public domain.

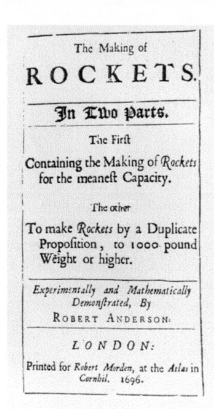

Figure 1-2   *The Making of Rockets by Robert Anderson. Image in the public domain.*

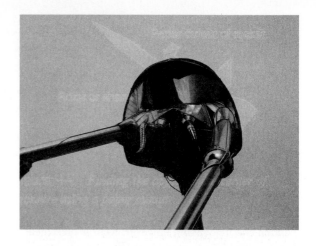

Figure 1-3   *Tripod statue in Woking town centre, UK. Image in the public domain.*

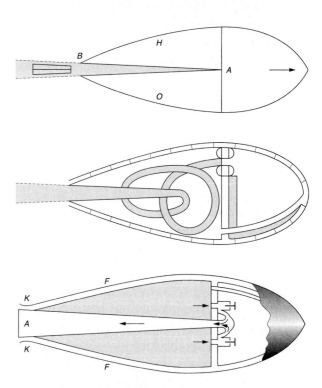

Figure 1-4   *Some of Tsiolkovsky's rocket designs. Image courtesy NASA.*

with space, and the possibilities of extraterrestrial beings from other planets.

Incidentally, a statue of one of the tripods depicted in the book was unveiled in Woking town centre to commemorate the 100th anniversary of the book (see Figure 1-3).

Space travel was discussed as early as 1903, when a Russian high school teacher, Konstantin Tsiolkovsky, published (ИССЛеДОВАНИе МИРОВЫХ ПРОСТРАнСТВ РеаКТИВНЫМИ ПРИбОРаМИ) *Exploration of Space Using Motors;* however, it was several decades before the technology to make this a reality was mature. He suggests the use of liquid hydrogen and liquid oxygen as an ideal propellant for use in rockets. You can explore this further in Project 4: Build Your Own Hydrogen-fuelled Rocket.

Robert Goddard is universally proclaimed as the father of modern rocketry. In 1917, the Smithsonian Institution gave him a grant to enable him to research rocketry. He attached a "de Laval" nozzle to a rocket

motor to greatly improve its efficiency. The "de Laval" nozzle is explored further in Chapter 2.

Peaceful development of rocketry continued, until in 1932 the German Reichswehr, later called the Wehrmacht, began to take an interest in the use of rocketry as a weapon. The Treaty of Versailles had curtailed Germany's military power, and left it with

Figure 1-5 *Robert Goddard and one of his rockets. Image in the public domain.*

Figure 1-7 *American "Redstone" rocket. Image courtesy NASA.*

Figure 1-6 *V2 shown thrusting away from the launch pad. Image courtesy NASA.*

very little in the way of long-range weaponry. The rocket was seen as a way of carrying explosive payloads over long distances to cause untold damage. The rocket made massive advances in size and range, and was named Vergeltungswaffe-2 – Vengeance Weapon 2 – known commonly by the abbreviation V2.

After World War II, United States, Russian, and British scientific and military organisations raced to grab technology and academics from Peenemünde where the German rocket program was based. As part of Operation Paperclip the United States was to get the lion's share of the resources, including many of the Nazi rocket scientists who worked for NASA in the early days of the space program.

The V2, originally a machine of war, was to evolve into the U.S. "Redstone" rocket used in the infant space program. Simultaneously, the Russians developed the V2 into the R1, R2, and R5 rockets.

In October 1957, the Soviet Union launched their satellite Sputnik (Спутник). Meaning "travelling friend"' the satellite started the space race – this event was to change so many variables in science and

Figure 1-8   *Sputnik.*

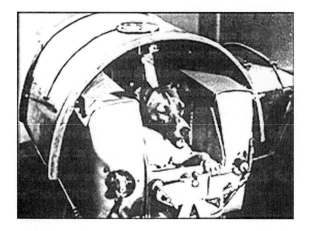

Figure 1-9   *Laika – the first dog in space.*

Figure 1-10   *Yuri Gagarin – the first man in space.*

Figure 1-11   *Landing on the moon.*

technology that the world has never been the same since. The competitiveness of two great nations has spurred on the development of so much innovation and invention that the world has truly benefited from man's foray into space.

From electronic payloads, the world quickly advanced to biological payloads. It was the Russians who first launched a live animal into space: Laika the dog. In this book, we will certainly NOT be launching any biological payloads!!

This biological payload paved the way for one of man's greatest achievements: escaping his homeland's atmosphere into the dark, empty void that is space. The Russian cosmonaut Yuri Gagarin was the first man in space; this served to intensify the space race further and led to one of the most bold and ambitious projects in the history of mankind. On July 20 1969, the US finally stole a march on Russia, landing Neil Armstrong and Buzz Aldrin on the moon.

The history of model rocketry is somewhat intertwined with that of the "real rockets." Looking at the size of Robert Goddard's original creations, it could be argued that they were nothing more than models; certainly it is possible for the model rocketeer to build much larger, more powerful, creations than those of the early days.

Model rocketry came about in 1954 when Orville Carlisle, and his brother Robert, set about the problem of designing small rocket motors for use in lectures on rocket propulsion. They followed the work of G. Harry Stine, in *Popular Mechanics* and in 1957 they sent him a sample of their rocket engines as Stine was a safety officer at the White Sands missile range. G. Harry Stine would go on to devise the original model rocket safety code, and to write *The Handbook of Model Rocketry*.

Many of the early model rockets were dangerous and blew up on the launch pad. The problem was the engines; because they were made by amateurs, they were not made to high tolerances.

The crucial leap came in 1958; Vern Estes had devised a machine to make model rocket motors. Now the motors were mass produced, they were made to high tolerances, and finally became safe and reliable. In 1960, Estes began selling motors through the post – model rocketry had finally come of age.

Many of the projects in this book have come about as the result of incremental developments and innovations by model rocketeers. Much has been done over the past several decades to develop the hobby to where it is today. It will be interesting to see what developments will result from the advances in the next few decades – where will the hobby go? It is down to you – the next generation of rocketeers – to determine that.

Chapter 2

Rocket Science

So you wanna call yourself a rocket scientist (Part 1)...

In order to be able to call yourself a rocket scientist, you have to learn rocket science. Many people perceive math, physics, and chemistry as dull. Unlike math, physics and chemistry are really interesting. This is why they are at the front of the book and the math is at the back! Personally, I find math a Necessary Evil; many will disagree, and you certainly can't ignore it if you want to be a rocket scientist. The safe execution of any model rocket project involves all of the components being calculated to work together in harmony to perform flawlessly.

Of course, this doesn't ALWAYS happen in real life; even NASA get it wrong sometimes, but before "launching" into building rockets in a foolhardy manner, it is worth considering the consequences of getting it wrong. At this level of rocket science, a few dollars, pounds, or euros are at stake ... There is no reason why this can't be interesting though ...

In this chapter, some of the principles of why rockets work are explained in an easy way supported by experiments. Rocket science isn't fantastically hard to understand, and you will certainly understand the concepts once you work through the experiments in this chapter. While the experiments aren't directly related to model rockets, they certainly are very useful; I would STRONGLY recommend working through them, whether for fun or not.

Let's start the chapter with a bang, the explosive project "Making Black Powder".

Project 1: Making Black Powder

Note

You should NEVER attempt to make your own rocket engines. This experiment serves as a demonstration. Trying to build your own rocket engine is irresponsible and dangerous. The rocket engines manufactured by the professionals are made to very high tolerances with very precise specifications.

This experiment is intended to be a stand-alone experiment to be carried out by a responsible adult who is familiar with chemistry. Only use VERY small quantities of the materials listed and NEVER store them in a closed container.

You read a little about black powder in the Chapter 1: The History of Rocketry. Arguably, never in the history of man has a chemical reaction had such a profound effect on society. Now we are going to make some in very small quantities in order to understand the chemistry behind model rocket engines and how they work.

You will need

- Charcoal
- Sulfur
- Potassium nitrate (saltpeter).

Where to get the chemicals

Charcoal is available in the form of charcoal briquettes for barbecues, or sticks of artist's charcoal. Sulfur is available from garden centers in the form of "sulfur candles." Potassium nitrate, saltpeter or nitrate of potash is available from chemical supply outlets.

Tools

- Mortar and pestle

How it is made

"True" black powder is made by mixing the three components together in the correct ratio. They are ground to a fine dust, which is highly explosive. Different ratios of the constituent components produce different burn characteristics. Mixing all three components and grinding them with a mortar and pestle is dangerous. I suggest that you grind the powders individually and combine them carefully later. While this does not make true black powder, the mixture demonstrates the effect and is less volatile; it is called pulverone. The ground powders should be mixed in the ratio 75 parts potassium nitrate:15 parts charcoal:10 parts sulfur.

To ensure your and others' safety, ignite any mixture remotely and make sure that you are standing some distance away with a protective screen between you and the mixture. Two things to note: first, "carefully," second, that the constituents should only be mixed with wooden, glass or stone implements, certainly no metal or anything that has the potential to spark.

How does it work?

Black powder burns very rapidly. The charcoal provides the fuel to burn, while the saltpeter (potassium nitrate) provides an oxidizer. All solid fuel rockets work on the same principle – they create a fuel that burns incredibly rapidly although it does not explode. This fuel produces a reaction mass which is forced through a de Laval nozzle. Solid propellent engines have the inherent advantages of being simple, cheap, and safe; however, they do not afford the same level of control that liquid rocket engines permit, and furthermore, once ignited they cannot be stopped.

Solid propellant rocket engines

Having made black powder, we can examine the way that a model rocket engine functions. Let's consider "rocket engines" in very general terms.

A rocket motor is generally an internal combustion heat engine. I say "generally," because some experimentation has been done with ion engines and other devices that use other phenomena to power the rocket. The rocket engines we are dealing with are called internal combustion heat engines because they burn fuel internally, and produce a reaction mass in the form of a hot gas. Figure 2-1 shows a selection of model rocket motors.

Figure 2-1 *A selection of model rocket motors.*

Model rocket motor construction

Let's look at the construction of a typical rocket engine. Refer to Figure 2-2 for more details. First of all, if you take a look at a model rocket engine, or failing that some of the pictures of rocket engines later in this book, you will see it is constructed within a rolled paper tube. The tube is made from a strong grade of engineering paper and has a number of functions:

- it contains the propellant,
- it supports the ceramic components of the rocket engine,
- it provides a uniform size to enable any engine from any manufacturer to work in the rocket.

Within this tube, working from the left, we first encounter a nozzle. This is specifically called a de Laval nozzle, and is discussed below.

The throat of the nozzle is the hole in the middle where the hot gas – which provides the thrust – comes out. This is also where we insert our igniter. The igniter makes contact with the propellant grain which is inside the rocket. It is this grain that provides the thrust as it burns. Once the main grain has burned, we move onto the smoke delay grain. This is a slower burning grain and produces little or no thrust. Its function is to allow the rocket to "coast" after it has been launched. Because the rocket gains momentum after its initial acceleration, in order to attain maximum altitude, it must be allowed to use up this momentum before the recovery system is deployed.

The reaction mass is made by combining a propellant with an oxidizer and combusting them within a chamber. This reaction mass is then forced out of a nozzle which allows it to expand very rapidly.

In the case of our model rocket engines, the propellant is a solid propellant which comprises a fuel and an oxidizing agent. As we explored in the last experiment, the charcoal is our fuel and the saltpeter the oxidizing element.

Some of the large rockets launched by NASA use liquid hydrogen and oxygen as their fuel. We will see a little more of that in Project 4: Build Your Own Hydrogen-Fuelled Rocket.

Now I mentioned that the reaction mass was forced out of a nozzle, known as a de Laval nozzle. Read on...

The de Laval nozzle

The de Laval nozzle was invented by an ingenious Swede by the name of Gustav de Laval in the 19th century. Some other sources might refer to this type of nozzle as a convergent–divergent nozzle, a con–di nozzle, or a CD nozzle, but essentially they all mean the same thing. The nozzle is shown diagrammatically in Figure 2-3.

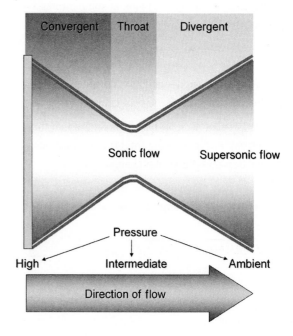

Figure 2-3 *The de Laval nozzle.*

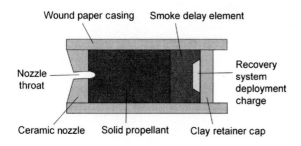

Figure 2-2 *Model rocket motor construction.*

You will see in the diagram the areas named convergent, throat, and divergent. Working from the convergent end, we see a gray rectangle. This rectangle represents the fact that beyond this point is a closed chamber. This is a combustion chamber where the fuel is burnt, and then forced by its own pressure into the convergent. In our model rocket engine, the combustion chamber is formed within the cardboard tube of the motor casing. The flame front grows as the combustion starts from the nozzle, and expands as the solid fuel burns. Some motors are called core burners; these have a hole drilled through a large portion of the propellant grain. This exposes a larger surface area upon ignition, and results in quick performance, albeit for a shorter duration. Drilling motors in order to increase performance is strongly advised against. Motors are made to very high tolerances and specifications by manufacturers, and tampering with the propellant grain of any model rocket motor greatly increases your chances of injury in what is a very safe hobby if executed correctly.

Returning to our de Laval nozzle, we can see clearly that the area labelled convergent is where the nozzle funnels in (converges). This is followed by the "throat" which forms a constriction in the neck of the model, followed by the "divergent," which is where supersonic flow occurs.

We will look at this in some more detail, but first let's define a few things and look at some impressive techy words! These words relate to the field of study called Thermodynamics – highly riveting stuff.

- Gas flow through a de Laval nozzle is adiabatic, this is to say, nearly no heat is lost.
- The process is isentropic. This means that the entropy of the working fluid (in this case our reaction mass) does not change.
- If you don't already know, entropy means the embodied heat energy that is *not* able to do work for us.

In the nozzle, the gas, our reaction mass, is produced by the burning solid propellant. From here it is forced into the convergent part of the de Laval nozzle. During this time, the gas is flowing at less than the speed of sound, and is compressible. The speed of the gas must increase as it passes through the constriction, as our mass flow rate – the rate at which our reaction mass flows – is constant. As the reaction mass enters the throat, the gas goes from less than the speed of sound, subsonic, to be transonic, Mach 1. Once it passes this point, the gas goes into the divergent, where it is allowed to expand. As it expands, it accelerates, increasing the amount of thrust produced, and hence making rockets possible.

Real rocket engines – the Bernoulli effect

Another concept to get to grips with is the Bernoulli effect. Bernoulli's principle states that an increase in velocity results in a decrease in pressure. This helps to explain the changes in flow rate as the combustion products flow through the de Laval nozzle.

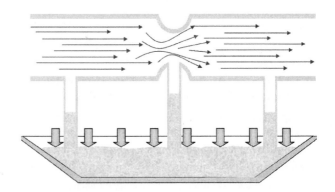

Figure 2-4   *The Bernoulli effect.*

Project 2: Demonstrating Different Rates of Reaction with a Film Can Rocket

You will need

- Four 35 mm film cans (same brand; translucent preferable)
- Alka-Seltzer
- Denture cleaner
- Effervescent vitamin C
- Vinegar
- Baking powder or bicarbonate of soda
- Water

Tools

- Scales
- Timer
- Measuring cylinder

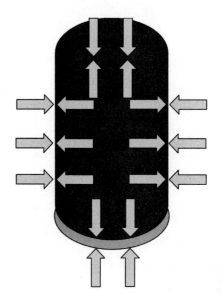

Figure 2-5i *Balanced forces: nothing happens.*

This project experiments with different rates of reaction, and looks at how the rate of reaction affects the thrust generated by a chemical rocket engine. By the end of this project, you should understand why a fast chemical reaction is needed in order to generate a lot of thrust and propel our model rocket to great heights.

Simply, we are going to take a film can and insert a measured amount of "propellant." We are going to look at some different chemical reactions and see which ones react vigorously and which ones react slowly. The chemical reaction will generate pressure inside the sealed film canister. This pressure will exert a force on the film canister as shown in Figure 2-5i; as can be seen, the forces are balanced – as a result nothing happens. When that force crosses a threshold, the lid of the canister will pop off and the canister will fly into the air. This is because the forces cease to be balanced. This is illustrated diagrammatically in Figure 2-5ii.

As with all good science experiments, this one can be a little unpredictable at times!

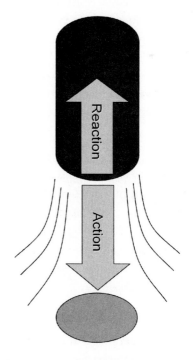

Figure 2-5ii *Unbalanced forces: rocket flies into the air!*

Isaac Newton's Laws of Motion

Isaac Newton was a clever chap. He can be forgiven his dubious hairstyling because he did many great things for the world, in particular advancing the sciences.

Godfrey Kneller's portrait of Isaac Newton (1689) oil oncanvas.

His three laws of motion can be used to help us understand the film can rocket experiment and also our hydrogen pipette rocket. His three laws of motion a were written in Latin in 1687 – the universal language of science at that time!

Newton's Laws of motion are written thus:

Axiomata sive leges motus

Lex I

Corpus omne perseverare in statu suo quiescendi vel movendi uniformiter in directum, nisi quatenus illud a viribus impressis cogitur statum suum mutare.

Lex II

Mutationem motus proportionalem esse vi motrici impressae, et fieri secundum lineam rectam qua vis ilia imprimitur.

Lex III

Actioni contrariam semper et aequalem esse reactionem: sive corporum duorum actiones in se mutuo semper esse aequales et in partes contrarias dirigi.

Doesn't make much sense?

What this means in English is:

First law:

Objects in motion tend to stay in motion, and objects at rest tend to stay at rest unless an outside force acts upon them.

Second law:

The net force on an object is equal to the product of its mass and its acceleration.

Third law:

To every action (force applied) there is an equal and opposite reaction (equal force applied in the opposite direction).

When your rocket is sitting on the launch pad, the only force acting against it is gravity. It is prevented from falling by the physical limitations imposed by the ground and the launch pad, which stop it from falling further towards the earth.

If we don't move the rocket, the rocket won't move itself – this is what the first law says: "Objects at rest tend to stay at rest," then when we fire the rocket motor the rocket gets underway. The other part of the first law says that "Objects in motion tend to stay in motion," but hang about, our rocket eventually slows and stops – this must mean that there are other forces acting on the rocket and in fact there are: drag and gravity discussed later in this chapter.

The second law says that the net force on an object is equal to the product of its mass times its acceleration ($F=AM$). My physics teacher Mr Spary used to say "forces are magic" as a method of remembering this formula. When you think about how this applies to rocketry, the amount of force that our rocket motor needs to produce is a product of how heavy our rocket is, and how fast we want it to go. This makes sense!

Finally, the third law says that to every action there is an equal and opposite reaction – this means that if our rocket motor is thrusting down, there must be an equal force pushing up against it. When forces are balanced, nothing happens; however, by unbalancing the forces, we can make things happen.

Figure 2-6 *Household vinegar; baking powder; effervescent vitamin C; Alka-Seltzer; denture cleaner.*

In addition to looking at the forces at work, we are also going to look at the rate of reaction, and how the rate of reaction affects the length of time, and the amount of power our rocket has to launch.

In order to conduct a fair test, weigh the tablets of the Alka-Seltzer, denture cleaner, and vitamin C, and take the value of the lowest one; cut pieces off the other tablets until they all weigh the same. Now measure out the same amount of the baking powder or bicarbonate of soda. The reagents are pictured in Figure 2-6.

In the first canister place the denture cleaner, in the second the effervescent vitamin C, in the third place the Alka-Seltzer, and in the final one place the baking powder/bicarbonate of soda.

The first three pots will require water to initiate the reaction, the baking powder/bicarbonate of soda requires vinegar.

One by one, take the pots and add a measured amount of liquid. Experiment with different amounts of liquid, but whatever you do ensure that for each test all four pots receive the same amount of water/vinegar.

Once you have added the liquid, put the lid on very quickly and start the timer. Stop the timer when the film can "pops" and flies into the air. Note how high it flew.

You should have been able to see the reaction progressing inside the film can. Was it a vigorous reaction or a slow one? How high did the film canister fly? How long did it take to launch?

By evaluating all of these questions for each batch of chemicals, you should be able to see that a fast chemical reaction is essential in order to generate the quick burst of thrust needed to get a rocket into the air.

How does this relate to model rocket engines?

The propellants in a model rocket engine burn INCREDIBLY fast; the whole motor is expended in a matter of seconds. This reaction must occur quickly in order to generate the volume of "push" required to get the rocket into the air.

Furthermore, you should be able to see that our propellants generated gas as an exhaust. Rocket engines need to produce a gas, in order to produce the massive *volume* of reaction mass from such a small amount of propellant. If rocket engines produced liquid for their exhaust, then the weight of the rocket would be so heavy to start with, they wouldn't be able to lift off very far at all.

If you want to demonstrate that the products of the above chemical reactions are all gases, then take some of the chemicals, mix them in a small soda bottle, and stick a balloon over the neck of the bottle – you will see the balloon inflate with a gas – this is the product of the chemical reaction.

Taking it further...

An additional experiment that you could perform is to look at how the surface area of the tablet reagents relates to the amount of time they take to produce the thrust. You might like to compare the same weight of crushed-up tablet to whole tablet. In some instances, you will find that the crushed-up tablet reacts so fast, it is hard to put the lid on against the volume of gas being generated to provide thrust.

How does this relate to model rocket engines?

The rate at which the propellant burns in a model rocket engine is determined by the amount of propellant exposed to the flame front. Some engines are called "end burners" – in these, there is a small amount of propellant exposed to the flame front, and they burn from one end to the other. This is analogous to the whole tablet, which emits gas steadily but over a longer period of time. Other rocket engines are known as "core burners." Essentially, they have a long hole drilled through their centre, with a large amount of propellant exposed to the flame front. As there is a higher surface area, the reaction takes place faster. This results in a quicker boost of energy accelerating the rocket at a greater rate. The core burner rocket motor is analogous to the crushed tablet.

Project 3: How Does the Quantity of Propellant Affect the Distance Travelled?

You will need

- 3 balloons

This experiment may seem simple, almost intuitive, but it is certainly worth doing just to reaffirm your understanding of the laws of physics, and furthermore, makes a nice simple demonstration for a science fair or classroom talk.

Take three balloons. Fill the first up with lots of air, half-fill the second, and give the final one a just a tiny squirt of air. Now hold them and release them.

It can be seen that the rocket with the most propellant, in this case our first balloon, produces the largest volume of reaction mass and as a result flies further and faster. The next rocket, which contains less propellant produces less reaction mass, and so travels a smaller distance. Finally, the rocket with only a little reaction mass does not travel far at all. This is analogous with rocket engines. Rockets with a lot of propellant travel very far indeed! Compare the volume of one of the space shuttle's boosters with the volume of a 1/2A rocket motor and it quickly becomes apparent that there are very real differences in the amount of propellant and the capabilities of the rockets.

Project 4: Real Rockets! Build Your Own Hydrogen-fuelled Rocket

You will need

- Manganese (IV) oxide
- Zinc scraps
- Hydrochloric acid (4.0 mol l^{-1})
- Hydrogen peroxide (3%)
- Two boiling tubes
- Gloves
- Safety spectacles
- Microbore tubing
- Two rubber bungs
- Cork lid
- Paperclip
- Plastic pipettes

Tools

- Tesla coil
- Awl

Making a suitable bung

Take a rubber bung that fits your boiling tubes, and using an awl or a small drill make a small hole in it, through which you can insert some microbore nylon tubing. You will use this bung to help you collect the gases generated by the chemical reactions.

Producing oxygen

On the tip of a spatula, take a small quantity of manganese (IV) oxide and place it in a boiling tube. Now add 3% hydrogen peroxide. You should have about 2 inches or so of fluid in the bottom of the tube. Now quickly stopper the tube with a bung, which will have a small length of tube attached. Insert the tube into the pipette.

Producing hydrogen

Take a clean boiling tube, and drop in a couple of small pieces of zinc and a little hydrochloric acid so that there is about 2 inches of fluid in the tube. The hydrogen gas will bubble. Using a similar bung to that described above, you will be able to displace some of the water in the pipette.

To light your hydrogen-fuelled rocket you will need a Tesla coil; details of building one of these is covered in the great publication *Electronic Gadgets for the Evil Genius.*

Building a launch pad

You need a small rod to launch your hydrogen rocket. Take a small paperclip, straighten it out, and stick it into a large cork lid from a jar. The pipette can then slide onto the rod.

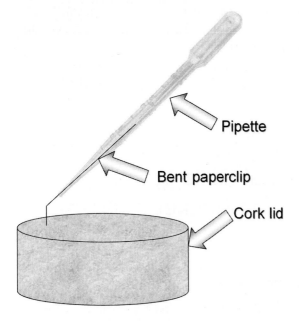

Figure 2-7 *Hydrogen-fuelled rocket launch pad.*

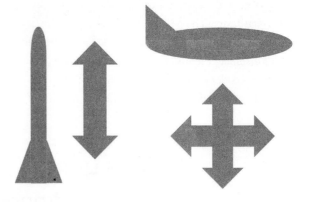

Figure 2-8 *Forces acting on a rocket compared to an aeroplane.*

Preparing for launch

Using the two boiling tubes full of hydrogen and oxygen, fill a small pipette half with hydrogen and half with oxygen. Leave about an inch of water in the bottom. This helps to seal the gases inside the tube, and furthermore provides a reaction mass to produce some meaningful thrust.

Launching the hydrogen-fuelled rocket

Hold the sparking end of the Tesla coil near to the base of the pipette where the gas meets the water. After several sparks, you will hear a loud pop, and your pipette will fly off the end of the paperclip, soaring into the sunset! Well, not quite, but at least over the other side of the room!

How does this relate to real rocket science?

The chemical reaction that is taking place here is exactly the same as the one that takes place in real rocket science. Many of the larger rockets that do not use solid propellants instead use hydrogen and oxygen in liquid form.

Forces acting on a rocket

We have already explored thrust in detail, but let's take a step back and look at the forces acting on a model rocket and compare them to an aeroplane.

Thrust, we know, is the upwards force that pushes our rocket skyward against the force of gravity. The weight of the rocket is acted on by gravity, which tries to pull our rocket back down to earth.

Now, what distinguishes our rocket from an aeroplane is lift. Lift can be defined as the force created by a wing or aerodynamic surface moving through a fluid, producing a force that lifts a body against the force of gravity.

Comparing an aeroplane to a model rocket (Figure 2-8), we can see that the rocket has no aerodynamic surfaces that are producing a force acting against gravity. The fins produce a "restorative force" by acting on the air, but we can't really call this lift in vertical flight; any lift produced by an errant model rocket flying horizontally is likely to be marginal. This is why model rockets are always launched vertically; the fins do not act as aerodynamic surfaces capable of supporting the weight of the rocket. This leads nicely onto our final force – drag.

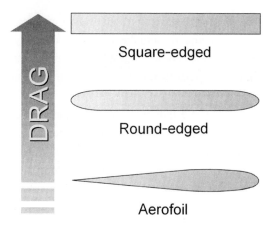

Figure 2-9   *How fin profile affects drag.*

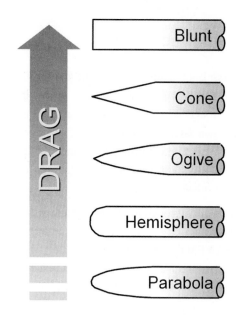

Figure 2-10   *How nose cone profile affects drag.*

Drag and rockets

The force known as drag slows our rockets down against the acceleration created by the motor. Drag is the force of the fluid that is the atmosphere, slowing our model rocket. Luckily there is a lot that you can do to combat drag, enabling your rocket to soar higher and faster.

One of the areas that we can concentrate on is the fins – a square-edged profile fin creates a lot of drag and turbulence. We can combat this to some extent by lightly rounding our fins front and back; however, the best fin profile for optimum performance is the tapered aerofoil fin, with a slightly rounded front, tapering back to a "knife edge." Figure 2-9 provides a quick at-a-glance reference.

As your rocket flies through the air, the first thing that hits the air is the nose cone. It is clear therefore that the profile of the nose cone *very much* affects the drag of the model rocket. Flying without a nose cone results in very high drag, as does flying with a blunt end. Interestingly, flying with a conical pointed nose is not the best shape for reducing drag. An ogive, with a slightly rounded profile, is better, a hemisphere better still, with the ideal rocket nose cone shape at subsonic speeds being a parabola. Figure 2-10 provides a quick at-a-glance reference.

The size of the body tube, very much influences the amount of drag. A larger body tube means a larger frontal surface area which entails higher drag. The thinner the body tube diameter, the less the drag. Of

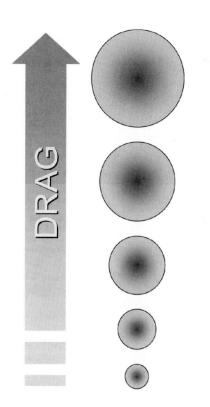

Figure 2-11   *How body tube diameter affects drag.*

course, there are limitations to how narrow you can make the tube – the obvious limitations will be the size of motor you intend to use in the tube and any payloads you wish to carry. See Figure 2-11.

Drag and recovery systems

While drag is a bad thing with regard to rockets, drag with recovery systems is a pretty good thing. Once our rocket reaches its peak (apogee) we need to bring it down safely to earth at a sensible speed. Too fast, and it will come hurtling down and probably embed itself in the ground and break (at best) or hit someone (at worst).

Rockets are designed to be sleek and aerodynamic – this is why with a comparatively small motor, they can achieve AWESOME heights. Unfortunately, once they have attained these awesome heights, they can be equally sleek and aerodynamic on the way down – not good!

What we need to do is find some way of making them not aerodynamic, and this is where a recovery system comes in. In order to protect your investment (and unwitting passers-by) it's necessary to find some way of recovering your rocket by slowing it down (creating drag) as it falls from the sky to earth. You wouldn't jump out of a plane without a parachute, so why launch a rocket without one? Both need to be brought back down to earth. Recovery Systems are covered in Chapter 6.

Further reading

Iannini RE. *Electronic Gadgets for the Evil Genius*. McGraw-Hill/TAB Electronics, 2004.

Chapter 3

The Model Rocketeer's Workshop

In order to build, test, and fly model rockets, it is essential to have a well-set-up workshop with the correct tools and equipment. This chapter reviews some of the essentials, and looks at equipment that can be built relatively easily and cheaply. We will discuss a number of tools that you can build to enable you to work with rockets more easily. We will then discuss some of the tools that really are essential if you want to go into model rocketry as a serious hobby. This will culminate in the construction of a wind tunnel, which is a very useful tool for looking at model rocket aerodynamics, and certainly provides a lot of room for experimentation for science fair or physics projects.

Project 5: Build a Model Rocket Support Jig and Stand

You will need

- Sheet of plywood/MDF/similar constructional material 4.5 × 16 in (19 × 40 cm)
- Length of 13 mm/18 mm/22 mm dowel
- Length of 2 × 2 in (5 × 5 cm) timber, to suit
- Panel pins

Tools

- Pin hammer
- Tenon saw
- Sanding block

In the following project, you are going to build a very simple jig to help you work on model rockets and a simple stand to allow you to support the rockets while painting.

Cylindrical tubes are awkward to work with on a bench. They crush very easily if clamped or put in a vice. Trying to cut a cylindrical tube leaning on a flat surface can be very dangerous, as there is a lot of potential for the tube to roll and move; and if you are using a sharp scalpel, there is potential for injury.

To remedy this, build a simple jig. The plans are shown in Figure 3-1. The jig is made from a piece of sheet material such as plywood or MDF. If you have the facility to work with acrylic, you could make this item as a single piece from a sheet of acrylic, creating the 90° bends with a line bender.

The finished article is shown in Figure 3-2. By holding a scalpel on top of the tube, and gently rotating it, you can effectively cut the tube without crushing it or making a bad cut.

The other tool consists of a number of lengths of dowel set into a piece of 2 × 2″ (5 × 5 cm) timber or similar. The diameters of the dowels correspond to the diameter of standard model rocket motor sizes. In order to support a model while painting, the dowel is placed in the tube conventionally used to hold the rocket motor. The rocket can then be painted. The model rocket support stand is shown in Figure 3-3.

What tools will I need?

The following is a list of tools that you should invest in if you are considering model rocketry as a hobby.

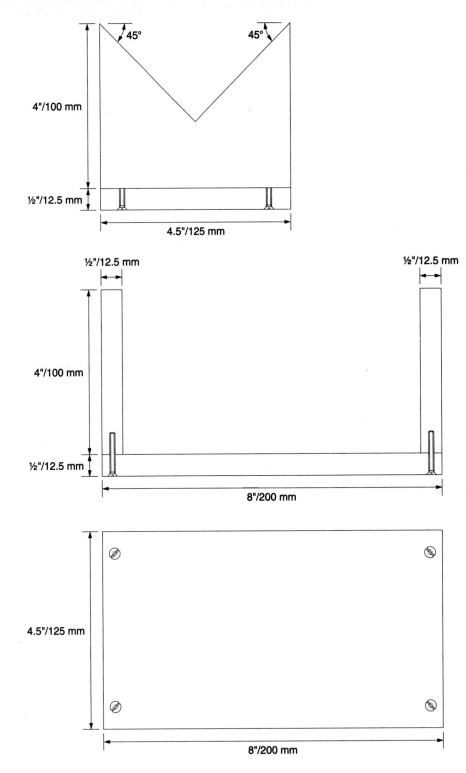

Figure 3-1 *Model rocket support jig construction plans.*

Most of them you will have already; remember it is only paper and cardboard you are working with, so in many cases, cheap tools will work well – nothing heavy duty is required, the only things that will wear out on a regular basis are sharp knife blades!

Knives and scalpels

It is essential that you have a good selection of sharp knives and scalpels when working with model rockets.

Figure 3-2   *Model Rocket Support Jig.*

Figure 3-3   *Model Rocket Support Stand.*

You will use them to cut card body tubes, balsa fins, sprue from plastic, and to do all other manner of jobs. Ensure that you use a knife with a good, sharp blade; using a knife with a dull blade can be dangerouss, as you apply more pressure to cut, and if you slip the results could be nasty.

Modellers' scalpels are fine for most jobs; for heavier jobs you may want to employ something a little meatier. Stanley brand knives are tried and tested and preferred by many people.

Have a peek at Figure 3-4. The top picture is of a disposable scalpel. While not shown in the picture for clarity, these come with a removable plastic safety cover which is removed for use. You should ALWAYS store these knives with the plastic safety cover on top.

When using a scalpel, it is always better to take several fine measured cuts, rather than a large gouge out of the workpiece. Hold your scalpel correctly like a pencil.

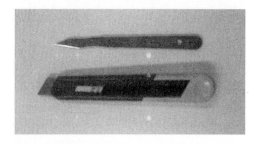

Figure 3-4   *Knives suitable for Model Rocketry.*

Figure 3-5   *Steel safety rule.*

The second knife is a disposable blade knife, where the blade is stored within the handle of the knife. The blades can be snapped off once they become too blunt, leaving a new sharp blade behind.

If you are a young rocketeer, you should seek an adult's advice before using knives such as these. There is no shame in asking for a little help – it is better to ask for help than to injure yourself severely.

Steel rule

DO NOT ever try to cut a straight line using a plastic ruler or other lightweight ruler intended for geometry or marking lines. The material is soft and is easily cut through by a sharp knife. Because of the plastic rule's low profile, it is also leaves the knife open to slipping and slicing your fingers. Even if you don't end up hurting yourself, the scalpel will tend to cut into the rule from time to time, nicking it and causing any lines that are drawn in the future to have bumps! A scalpel will go through flesh VERY easily; bear this in mind when debating whether to spend a few dollars more on a safety rule or not.

Instead of risking hurting yourself by using a plain plastic rule, use a steel safety rule, such as that shown in Figure 3-5. The rule has an elevated central section; this allows your fingers to hold the rule safely out of the way. Furthermore, the scalpel won't cut through stainless steel and so will follow the line without deviation or injury.

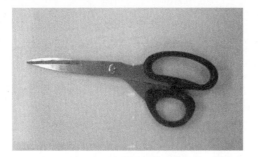

Figure 3-6   *A sturdy pair of scissors.*

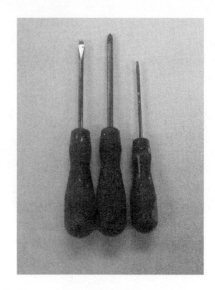

Figure 3-8   *A useful set of screwdrivers.*

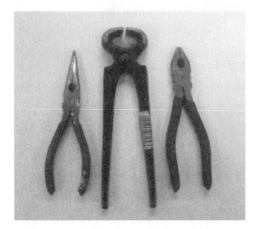

Figure 3-7   *A selection of pliers and pincers.*

Figure 3-9   *A set of "Tank Cutters" (hole saws).*

Scissors

A small pair of scissors such as that found in a nail/manicure set will also be incredibly handy from time to time for those small awkward fiddly jobs. In addition, you will find a larger set of scissors very useful for tasks such as cutting out parachutes and streamers and cardboard parts. For the U.F.O. projects in this book, scissors are a must!

Pliers, pincers, and grips

Pliers and pincers find many uses when manipulating small parts. There are some jobs that are simply too fiddly to perform or hold with fingers. Another application where a pair of long-nose pliers finds many friends is when gluing things with superglue. Far better to get glue on pliers than on your hands.

Screwdrivers

It goes without saying that you should have a selection of screwdrivers. For some of the electronic payload projects you will need some fine jeweller's screwdrivers to enable you to take apart a digital camera.

Hole saws

While their use may not be immediately apparent, I can highly recommend a set of hole saws, commonly known as tank cutters. In addition, as you will most likely be working with lightweight materials, investing in a trammel is useful for those odd-sized holes that aren't covered by your hole cutter set. As these will be mostly used to cut balsa and thin ply, even a relatively cheap

Figure 3-10   *A wood lathe – shown turning a nosecone.*

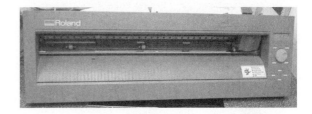

Figure 3.11   *Roland CAMM-1 Vinyl Cutting Machine.*

set will prove useful and long lived. The reason I recommend their purchase is that you will find them invaluable for making your own bulkheads, motor spacers, and custom multiple motor mounts – items that are relatively expensive to purchase considering the cost of raw materials. If you can pick sizes that correlate carefully to the sizes of model rocket tubes and motors you will be using, then you are certainly onto a winner!

Lathe

A small lathe is a useful investment if you intend to make many model rockets. Use it to turn nose cones – while small nose cones will give you change out of a couple of dollars, anything over two and a half inches will not leave you with much change from a $10 bill, and for larger, high-powered rockets, noses costing $30 and above are not uncommon.

Vinyl-cutting machine

What follows is a real luxury that is likely to be well out of the reach of the average rocketeer; even so, if you are aware of its usefulness, you are sure to find devious ways to procure access to one. Failing that you could find a good second-hand bargain in an online auction – I have seen working models sell for around the $100 mark.

Vinyl cutting machines are used by signwriters to put signs on storefronts and on the side of vehicles. They accept a sheet of vinyl which is backed onto a glossy

paper substrate. The co-ordinates of the desired legend is fed to the signwriter from a computer running a vector graphics programme. The signwriter then plots this pattern onto a sheet of the vinyl using a small cutter with a sharp tip. This cuts the vinyl, but the pressure is insufficient to cut through to the paper. Once the design has been cut by the signwriting machine, a scalpel can carefully be used to remove the unwanted vinyl. The finished design can then be lifted from the support paper using Frisk tape or a low-tack adhesive tape and transferred onto the sign, or in our case, model rocket.

The advantage of using a vinyl signwriter is that the amateur rocketeer can produce clean, crisp decals that will happily compete with those produced by any professional manufacturer. Furthermore, because they are cut out of bright durable vinyl, you have vivid colours that will not fade quickly.

Soldering iron

Some of the later projects in the book will certainly require a soldering iron and solder. You need a soldering iron that is suitable for electronics work – if you use a soldering iron designed for larger items of metalwork, you may find that you fry the delicate components you are working with!

What sundries will I need?

Most simple model rocket kits you can simply get out of the box and almost snap together; however, some of the slightly more advanced model rockets will make increased demands upon the hobbyist. A slightly larger array of sundries will be required for quality assembly

Figure 3-12   *Soldering iron and solder.*

Figure 3-13   *Cyanoacrylate adhesive.*

of the model. These are the sort of things you don't want to skimp on, because they are relatively cheap, and quality construction can often mean the difference between a model that you fly once or twice, and a model that can be re-used many times.

Glues

Cyanoacrylate – Super Glue

Cyanoacrylate adhesive (Figure 3-13), commonly known by the brand name Super Glue is ideal where jobs need to be done rapidly! This glue tends to be a little brittle when dry; however, it proves very useful as it does not necessitate holding or clamping items for a long amount of time. For really fast gluing, cyanoacrylate adhesives can be used with a spray activator that allows the glue to set even faster!

Balsa cement

For models such as boost gliders, balsa cement (Figure 3-14) is a must-have item. It joins together lightweight balsa parts quickly and without worry. Because balsa is so light, clamping items in place while the glue dries is rarely a worry; any positioning or holding can be done with a bit of masking tape.

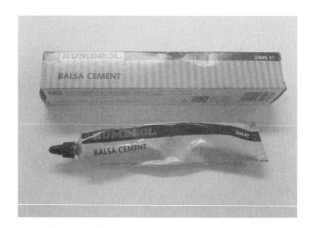

Figure 3-14   *Balsa cement.*

Poly cement

There are many instances in model rocketry where plastic parts need to be joined, a common example would be many nose cones which come as a two-part assembly – the nose itself and the shock cord mount. Ensure that you have some poly cement to hand to glue nose cones and other plastic parts.

Clear glue

Clear glue (Figure 3-15) finds many friends in the model rocketry world. Recently there have been moves to replace solvent-based clear glues with solvent-free

Figure 3-15 *Clear solvent-based glue.*

Figure 3-16 *Selection of tapes.*

products; some hobbyists complain that the solvent-free products do not perform as well as the equivalent solvent-based products.

White glue

White glue, also called PVA or polyvinylacrylate adhesive is a very useful glue for model makers. It start out life as a white runny glue, but sets to a translucent plastic-like finish.

Tapes

Adhesive tapes find no end of uses in model rocketry. Whether it be fixing a body tube, holding something in place temporarily, or jerry-rigging payload electronics into their position at the launch site, a good selection of tapes will hold you in good stead.

Paints

An airblown finish when executed correctly looks much much better than its streaky handpainted counterpart. A wide range of spray paints is available from hobby stores and paint factors in a many colors and finishes. Your imagination is the limit when it comes to model rocketry, models can range from being finished in flat

Figure 3-17 *Cans of spray paint and lacquer.*

colours to metallic colors, and even for the skilled amateur "candy" colors and "chameleon" colors that change depending on the angle they are viewed from.

Plastikote are my favourite brand of paints, they sell cans which are small enough for model-making purposes in a wide range of colors and finishes. Their chrome paint gives a particularly pleasing effect on cardboard – simulating real metal.

Painting

The thing that takes the longest with painting anything is the preparation. It doesn't matter if you buy the best paint in the world, the surface underneath is going to make or break the model. You really should spend the bulk of your time sanding, filling, and filleting the entire rocket to make sure the surface is silky smooth before you even look at a spray can.

Balsa   Any parts made of balsa should end up having a plastic-like finish before you attempt to paint them. Apply filler as above, and sand until the surface feels silky smooth. If you are a perfectionist, keep an array of sandpapers and work from coarse to smooth as you progressively give the surface a finer finish.

Plastic   Plastic can be tricky to paint as some paints can react with plastics, causing them to bubble or soften and deform. The old adage is to try the paint on an unseen area first of all, but taking this a step further, most plastic modelling parts come attached to some scrap plastic as a result of the process which produced them (injection moulding is fairly common). This scrap plastic called "sprue" is an ideal way of testing a variety of paints before applying them to your model.

A friendly model shop owner should be able to advise as to the compatibility of the paints they sell with various types of plastic used in model making.

When painting plastic, remember that your skin has a certain amount of natural oil on its surface, and as plastic is non-porous, anything transferred from your skin onto the part will stay on the surface, and could react or cause the paint not to adhere.

Before you start to paint, make sure that you clean the parts you intend to paint and wipe them thoroughly. A suitable solvent can be used to remove any grease or grime accumulated from handling and normal use. Again, test this first of all on an inconspicuous piece of plastic. Wait for the piece to dry thoroughly before beginning to paint.

This guide is intended as a supplement to manufacturers' instructions, not a replacement, so be sure to read them first of all.

- Shake the spray can in accordance with the manufacturer's instructions.

- Now, rather than spraying the part to be painted, check the spray on a spare piece of card or paper. (NO NO NO! Not on the cover of this book ☺) The spray pattern should be clear and free from spattering. If the nozzle appears to be clogged,

rectify the problem by changing the nozzle or cleaning it using a solvent that is compatible with the paint and/or a small pin or needle.

- When a satisfactory spray pattern has been achieved, you can begin to think about painting the model.

- When you start to spray, do not aim the can at the model, instead aim to the left or the right of the model. This is because the first bit of paint out of the can almost certainly comes faster than the rest, and the uniformity of the spray pattern cannot be guaranteed.

- Hint: It is far better to apply many thin coats and build up the finish rather than a thick coat which will tend to run and look unprofessional.

Pin-striping

Pin-striping is another weapon in the model rocketeer's war against blandness and search for individuality!

The addition of one or more self-adhesive coloured lines to your model rocket will result in it looking much more attractive than a plain finish, when executed tastefully.

Pin-striping is commonly sold by auto factors for adding a touch of finesse to your vehicle. As such it is generally sold in 10 m lengths to enable you to pin-stripe both sides. This provides us with more than ample pin-striping to do a number of model rockets!

Abrasives

A good selection of abrasives are essential in order to get a good finish on the parts of your model rocket. Carry a selection of different abrasives in your toolkit, from coarse grains for tidying up plastic parts, to finer grains for smoothing filler. When using abrasives, start with a coarse grade of paper, and work your way to a finer paper in order to achieve the best effect with minimal effort.

Signwriters' vinyl

For achieving a high-gloss professional finish, signwriters' vinyl is the presently unexploited hero of the model rocketeer (Figure 3-19). While it may be hard

Figure 3-18   *Selection of abrasives.*

Figure 3-19   *Signwriters' vinyl.*

to get the flat vinyl sheet to conform to shapes such as nose cones, signwriters' vinyl works terrifically for flat areas such as rolled around a body tube or cut to fit a transition piece. Instantly, the dull card surface is transformed to a shiny, weather-resistant, glossy, plastic-like finish, which will prove to be both, attractive and durable.

Fillers

Parts made from wood and balsa have an inherently "rough" surface (it may feel smooth to you, but aerodynamically it's *rough*). A little judicious filling will ensure that the part is smooth to the air, making your model fly more accurately and further, and also is visually attractive when painted.

Fillers can also be used to give a nice join between parts such as wings and the model rocket body. This not only looks attractive, but is also aerodynamically advantageous.

A little filler here and there can smooth the transition between model rocket parts, but remember, anything you add to the rocket carries a weight penalty which will have knock-on effects on the performance of the model.

Dope

Dope, in the model-maker's sense of the word, is the classic filler for filling and sealing surfaces. It comes in small bottles, and as with most paints and fillers, it is better to apply several light coats than one thick coat.

Figure 3-20   *Model Mate filler.*

Make sure that you sand lightly between each coat. This serves to "key" the surface and also to remove any "high-spots" which would only be exacerbated by a second coat of filler.

If you want to be really pernickety, you can even give the body tubes a light sand, and then apply a little filler to the spiral windings along the body tube.

Model Mate

One of my favourite fillers for model rocketry applications is manufactured by Carl Goldberg Models, and is called Model Mate. It is designed for use on model jet aeroplanes, it won't shrink causing your model to warp, and it is lightweight.

Polymorph

Polymorph is a material previously unheard of in the rocketry fraternity; once you have discovered it, you

Figure 3-21   *Polymorph.*

melted easily in a cupful of water. Once it has melted it turns clear and can be molded to any shape you desire. As it cools, it sets and hardens. The final result is a hard, tough opaque nylon-like plastic which can be machined and worked. It can be reheated and remolded; however, the final molded part has less of a surface area than the granules from which it was made, as such it is harder to heat and melt.

won't know where you were without it. Polymorph is a smart material, that allows small plastic parts to be made with relative ease and no expensive tooling.

Polymorph is a plastic with a low transition temperature. It comes in granulated form and can be

Online resources

www.hobbyplace.com/tools/xacsets.html

www.stanleytools.com

www.plastikote.com

www.stripeman.com/pin-striping.htm

www.carlgoldbergproducts.com

Project 6: Build a Wind Tunnel

You will need

- Flexible MDF
- Thin MDF sheet
- Semi-rigid transparent plastic sheet
- Panel pins
- Contact adhesive
- High-power desk fan

Tools

- Bandsaw
- Tenon saw

If you can get friendly with someone who works in the building services industry, or with sheet metal ductwork, the chances are you can build a *far* more sophisticated, durable, wind tunnel along these guidelines, using the zinc galvanized ductwork used in buildings for air conditioning.

We are going to be building a small tabletop wind tunnel in order to look at airflow over model rockets. This wind tunnel is not incredibly sophisticated, but it will allow us to conduct some interesting experiments at a basic level, and will serve as an interesting diagnostic tool.

You can make the tunnel's size to suit the materials that you have. The hardest part to procure is a clear plastic tube of a sufficient diameter. Look around Wal-Mart for something that is packaged in a clear plastic cylinder, often "display cases" of plastic from toys and the like can be customized.

If you look at Figure 3-22 the construction of the wind tunnel should be self-evident. The dimensions are based around that of the plastic pipe. From this, the flexible MDF was bent around and cut to form a cylinder. Within the MDF cylinder are baffles cut from thin MDF sheet. These help to smooth the airflow and create a smooth, laminar flow.

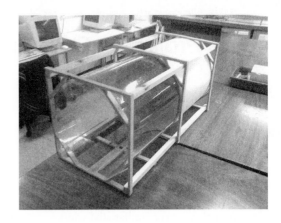

The rocket is supported in the clear section of the tunnel, while the fan draws air through the baffle. Hopefully, by the time the air reaches the model rocket, it is flowing smoothly and any turbulences and eddies induced by the fan are ironed out.

When supporting your model rocket in the wind tunnel, try to think of ways of doing so that will hold the rocket in place, without disrupting airflow. Use fine string or fishing wire to hold the rocket in place without disrupting the surrounding air currents.

In order to study the airflow, you need to use some flow visualization techniques. Move on to the next project for ideas that will enable you to see how the air moves around your creation.

Online resources

There is lots of information on the web about building small wind tunnels and a variety of experiments that can be performed with modest materials and equipment and on a small budget. The following links provide a springboard for your own exploration.

Franklin Institute's Guide to building a simple wind tunnel: sln.fi.edu/flights/first/makesimple/

http://quest.daps.arc.nasa.gov/aero/events/ collaborative/wind_tunnel.html

www.ceeo.tufts.edu/ldaps/htdocs/curriculum/ cheap_tunnel/winddir.html

www.ceeo.tufts.edu/ldaps/htdocs/curriculum/ cheap_tunnel/tunnel.html

www.science-projects.com/WindTunnel/WF htm

Figure 3-22 *Wind tunnel construction.*

Project 7: Flow Visualization Techniques

You will need

- Light cotton thread
- Pea stick
- Fine florist's wire (or similar stiff wire)
- Cyanoacrylate adhesive (Super Glue)
- Smoke pellets (used in central heating tests), or
- Spray Smoke (from theatrical retailers).

Tools

- Fine scissors
- Wire cutters

You can do some really cool stuff with a wind tunnel that allows you to visualize how air will flow over your model rocket while in flight. We can see the way that air interacts with the surface.

Cotton tufts

Using some low-tack tape such as masking tape, you can tape small lengths of cotton thread, say 2.5 cm/1 in or so long, to your model rocket body to see which way they blow in the wind. You can analyze their behaviour as the air flows over the rocket body. This can help you visualize airflow over your model.

Cotton probe

Take a thin dowel or pea stick, and affix a 4 in length of cotton to the end. This can then be used to "probe" specific areas of the model to see how the air is flowing.

Cotton grid

Take some stout wire such as florist's wire, and form a mesh. Use small dabs of cyanoacrylate (super glue) to join the intersections of the wire. Before the glue is fully set, poke the end of a short length of cotton into the intersection and wait for it to dry. Repeat this until each intersection has a small piece of cotton attached.

Troubleshooting

If the cotton or thread you use is too heavy it will tend to "droop" rather than pick up the airflow. Use a lighter thread.

Smoke

Introducing smoke into the airstreams is a very good way to visualize the flow forms around a model rocket. Smoke in quantity is easily obtainable. Speak to a friendly plumber, who will point you in the direction of "smoke matches," "smoke pellets," or a similar product. You may find that the smoke produced by some of these devices is excessive in quantity – some of them are designed to produce cubic metres of smoke in only a few minutes. If this is the case, break off smaller chunks of the pellets at a time. Be careful and take appropriate precautions, as you are dealing with flammable materials.

Troubleshooting

If the airspeed in the tunnel is too high, then the smoke will just be a blurry mess and you will not be able to get any meaningful information. In this case slow down the fans a little.

Surface oil flow visualization

Surface oil flow visualization is an experimental technique that helps engineers analyse the aerodynamic flow close to the body of an object. The technique allows you to analyse where flows separate and reattach. The method is quite simple. The body of the object to be tested is smeared with a thin film of oil. To make the flows easier to see, a pigment can be added to the oil. Because this technique can be a little messy, I would try it on an old model or a mock-up of your model rather than the real thing.

Chapter 4

Model Rocket Stability

It is important that our model rocket is stable in flight in order for take-off to be consistent and reliable When our rocket leaves the launch pad, it is a free body in the air, not supported by anything other than the thrust provided by its motors.

We saw how drag acting on a model rocket is one of the forces that opposes the thrust of the motor, but we need to consider a little more carefully how the air in the atmosphere acts on our model rocket in order to make sure that it is stable in flight.

A nice way of thinking about model rocket stability is to consider a weather vane as shown in Figure 4-1. We see that the weathervane has a "point" with a smaller cross-sectional area than its "tail." It is pivoted about the centre. Thus if a wind blows perpendicular to the weathervane, the tendency will be for the "tail" to move backwards. This is because there is a smaller area at the "point" of the weathervane than at the "tail," and so there is a smaller force acting on the point than the tail.

If we now think about a model rocket, rather than a weather vane, there are a number of key differences; first, our rocket is a three-dimensional object, rather than our flat weather vane, which could almost be said to be a flat plane. Substituting the rocket for the weathervane, the point of the weathervane is analogous to the nose cone of our model rocket, and the tail of the weathervane analogous to the fins of our rocket.

If we were to suspend our rocket in the air on a piece of stiff wire, and blow air from the side, we would find that the rocket would turn in the air like the weather vane. Now imagine moving the point of attachment point along the rocket and repeating the experiment. There would come a point where the forces balanced out and the rocket did not turn. This point is called the "center of pressure."

Second, our rocket is not suspended about a pivot like a weather vane; it moves freely in the air. However, there must be a "turning point." You will notice that if

you throw a pencil into the air, there is a tendency for it to flip about a point as it falls down; add a bit of blue tack to one end of the pencil and the point changes. You will see that as different objects are thrown into the air, they rotate about different points; that point will always be the same for any one object if its weight distribution doesn't change. This point is known as the "center of gravity."

The center of gravity is the balance point for weight distribution, and the center of pressure is the balance point for the aerodynamic forces that are acting on the model rocket.

Figure 4-2 shows two logos to denote center of gravity and center of pressure. These are fairly

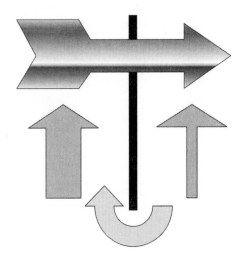

Figure 4-1   *The restorative force of fins.*

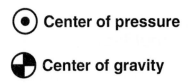

Center of pressure

Center of gravity

Figure 4-2   *Logos for center of pressure and center of gravity.*

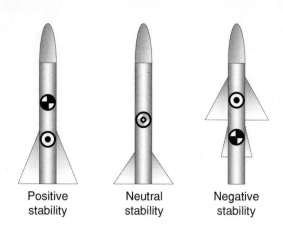

Positive   Neutral   Negative
stability   stability   stability

Figure 4-3 *Stability conditions.*

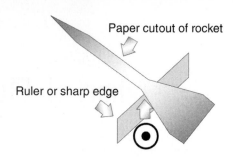

Paper cutout of rocket

Ruler or sharp edge

Figure 4-4 *Finding the approximate center of pressure using a paper cutout.*

standard and you will find them in many rocket books and kits.

The conditions of rocket stability are shown in Figure 4-3. In our ideal model rocket, we want the center of pressure to be behind the center of gravity by one body tube diameter or more. The pro's call this one-caliber stability, because caliber is another word for diameter of body. This condition is known as positive stability. It is a safe way to fly, and the way we want to go! Neutral stability is where the center of gravity and center of pressure are in the same place. Negative stability is when the center of pressure is in front of the center of gravity. This is not a good place to be!

Rocket stability diagnostic toolkit

The following are a couple of little tests that will get you on your way to building stable rockets. They are a useful tool to help you gauge various factors affecting the stability of your rocket.

Paper cutout test

A simple, practical way to try to determine the center of pressure of your model is to take some card, and draw a "profile" of your rocket on the card. This must be done

accurately, and it should represent a scale side-view of your rocket.

Now take this paper cutout, and a sharp edge such as a ruler or a razor blade. Balance the paper cutout on the ruler as shown in Figure 4-4. The point at which the cutout balances is roughly where the center of pressure will be.

Note: For the following tests, you will need to install your motors, or an equivalent weight to enable you to make accurate predictions.

String test

Finding the center of gravity of your rocket is also a relatively simple affair. Take a piece of string, and tie it around your rocket so that it is tight, but not so tight that the body is deformed in any way. You should be able to slide the string along the rocket body relatively easy. If the rocket tips tail downwards, then move the string back. If it tips nose down, then move the string further forwards. The point at which it balances is the center of gravity of your model (Figure 4-5).

Swing test

A swing test is a simple test we can do with our model rocket to assess its flight performance. It follows on

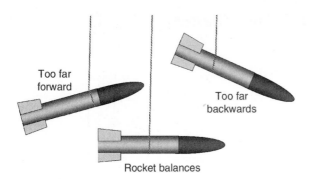

Figure 4-5 *The string test.*

Figure 4-6 *Swing testing a model rocket.*

logically from the string test, in that it requires that you have a piece of string tied around the center of gravity of your rocket (the point at which the rocket balanced; see Figure 4-6).

Now, holding your rocket, throw it and whirl it around you gently.

There are a number of things that can happen. If your rocket points in the direction of swing, and exhibits a slight wobble, then it is a good'un! This rocket will be safe and stable in flight.

If your rocket flies fins first, this could be because it is a very light model, with the center of gravity biased heavily towards the motor because of the relatively light weight of the rest of the rocket. It can also be an indication that the center of pressure is in front of the center of gravity.

If your rocket cartwheels, then this is a sign that it will not be stable in flight. You might want to add a bit of plasticine to the nose to smooth things out a little.

Gav's golden guidelines for great rocket design

- Increasing fin size moves the center of pressure backwards

- Adding weight to the nose or making the rocket longer moves the center of gravity forwards

Computer-aided design

The hobby of model rocketry has been totally revolutionized by the introduction of computers. From early days where model rocketeers would have to tediously key in lines of BASIC into simple computers with less memory than today's scientific calculators, the hobby has now evolved to provide fantastic CAD programs, which allow us to model many complex variables and provide data that can be interpreted easily.

There are two fantastic programs on the market that I could highly recommend to any model rocketeer; SpaceCAD and RocketSIM.

Both programs are relatively cheap and pack in a lot of functionality. At the time of publication Rocket SIM costs under $100 and you can buy SpaceCAD for under $60. When you consider the cost of materials that go into a new model rocket, you begin to realize that such programs are a good insurance policy for your investment.

Why would I want to use computer-aided design to design my models?

Well, if you are going to design models properly, you either have to use CAD software, or perform a lot of calculations manually. While it is good to understand

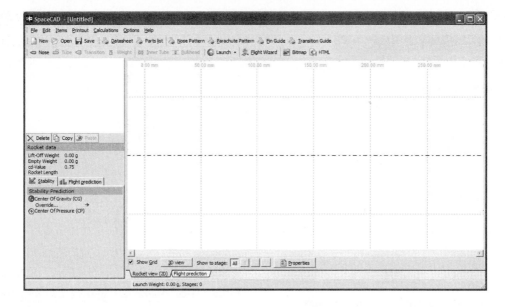

Figure 4-7 *SpaceCAD main window. Image courtesy SpaceCAD.com.*

what calculations and rules of thumb you can use, it is also immensely useful to use a computer in order to work out quickly and simply the many variables that can affect your model.

What other advantages are there to designing by PC? Well, for a start you will find that you waste less money. That is, less money in crashed rockets and less money in parts that you don't need. You see, when you design a rocket you will get a printout of exactly what you need from a standard library of components.

Also, there are features like printable templates that can be used for fins and parachutes.

Note: there are coupons in the back of this book for a free SpaceCAD demo CD!

Using SpaceCAD

Let's look at SpaceCAD and see how we can quickly put together a model rocket. First click on the SpaceCAD program icon.

Tip: If you have installed SpaceCAD with the default options, you will need to click on the following in Windows XP: Start Menu>>All Programs>>SpaceCAD>>SpaceCAD.

Once the program opens you will be presented with the main window, shown in Figure 4-7. You are

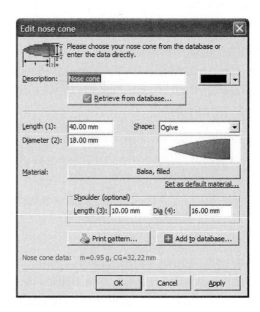

Figure 4-8 *SpaceCAD nose cone editor. Image courtesy SpaceCAD.com.*

presented with a series of little icons at the top of the drawing space. Starting from the nose and working our way down we will specify the components of our model rocket.

In the Space CAD nose cone editor (Figure 4-8) you can specify the type of nose cone that you are going to use in your model. If you retrieve the data from a

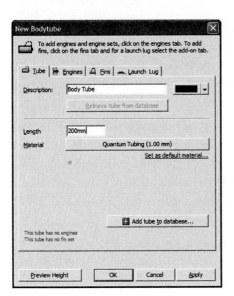

Figure 4-9 *SpaceCAD new bodytube tube editor. Image courtesy SpaceCAD.com.*

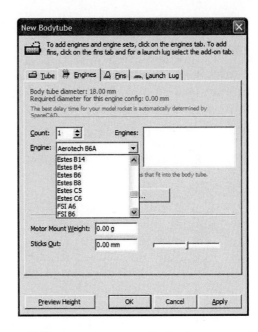

Figure 4-10 *SpaceCAD new bodytube engine editor. Image courtesy SpaceCAD.com.*

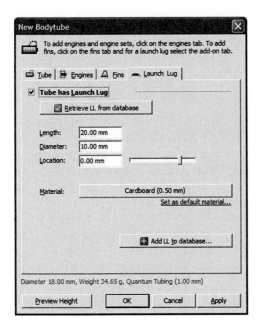

Figure 4-11 *SpaceCAD.com new bodytube fins editor. Image courtesy SpaceCAD.com.*

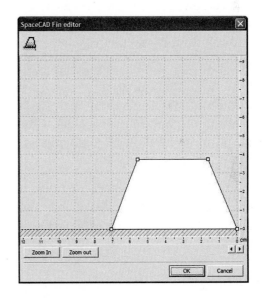

Figure 4-12 *SpaceCAD fin editor. Image courtesy SpaceCAD.com.*

database, you can use predefined settings for commonly available rocket nose cones. Failing that, you can enter the data manually for your nose cone specification.

Once you have done this click OK.

Next you need to specify the bodytube that you will be using (Figure 4-9). Aside from specifying the length and material, you will see that there are a number of

tabs at the top of the screen. These tabs are used to select different features that can be added to the basic tube. We will review these in turn.

The bodytube engine editor (Figure 4-10) does exactly what it says. It allows you to add an engine to your bodytube.

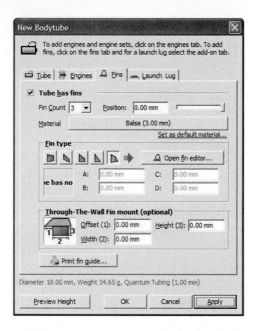

Figure 4-13 *SpaceCAD new bodytube launch lug editor. Image courtesy SpaceCAD.com.*

The bodytube fins editor (Figure 4-11) allows you to add fins to your bodytube. Specify the number of fins and the material they are made from. The SpaceCAD fin editor (Figure 4-12) allows you to draw fins for your model rocket by dragging out a shape using the handles.

Of course you are going to need to include a launch lug if you are intending to launch the rocket from a rod or a rail. Again, the SpaceCAD built-in database comes in very handy here as it allows you to paste in commercially available launch lugs (Figure 4-13).

Once you have worked through these options you have built your first SpaceCAD model! You are now well on your way to simulating model rockets with this software.

Chapter 5

Constructing Model Rockets

From top to toe – what you need to know!

Building model rockets isn't hard; building great model rockets isn't hard, but requires a little skill. You will get out of the hobby what you put into it. A sloppy, hastily constructed rocket will probably fly, but the chances are you will get more satisfaction from the model if you have put time and effort into making it yourself.

Similarly, it is possible to buy many components off-the-shelf, but many of these things can be made or improvized. Certainly, when you get into the realm of scale modelling, many components are not available to your exact specification and need to be fabricated.

We are going to spend some time looking at our model rocket from the top to the bottom, reviewing each step sequentially, and looking at how it is built. We will then follow up with some simple model rocket projects.

Nose cones

We took a look at optimum nose cone profile in Chapter 2: Rocket Science, and at optimum length-to-width ratio of nose cones in Chapter 4: Model Rocket Stability. Now we are going to investigate what they are actually made of, and how we can make them ourselves.

First of all we will look at the homebrew approach. Turning a balsa wood nose cone on a lathe really isn't a big deal. The material is so light that it cuts very easily; all you need is a set of sharp tools and a little patience. Often, you will need to do a little more hand-finishing "off-lathe," especially cleaning up where the center leaves its indentation; however, by and large you will find it easy and satisfying to make your own nose cones.

In my opinion, the hardest part of making a nose cone is turning the part to fit in the body tube. You want the transition between nose cone and body to be as smooth as possible with as little variation as possible. You also want the nose cone to fit snugly in the tube, but not so snugly that it will not eject as the ejection charge fires. You should take all this into consideration. The best approach is to take your time, go slowly, and if it does not work out right first time, have another crack at it.

If you don't have access to a lathe, it is possible to make something that will perform an admirable job of allowing you to turn balsa wood – it is very light after all. With an appropriate mandrel, some work can be done when turning balsa wood on a pillar drill; also, a rigidly mounted hand drill has proven successful for small jobs. However, if you are using an unorthodox method, be doubly careful about safety.

Of course, you need to find a way to fix your recovery system to your balsa wood nose cone. Go to the local hardware store, and buy some of the metal "eyes" that are used to affix net curtains. These only cost a few cents each and easily screw in by hand to the light balsa wood.

Of course, if turning your own nose cones seems like a lot of hard work, they can always be bought off-the-shelf. These are manufactured to fit all of the common body tube diameters. Some, such as the balsa wood nose cones shown, will benefit from a little time taken to fill and finish them (take a peek at Chapter 3: The Model Rocketeer's Workshop for advice on filling). Some plastic nose cones require a little finishing. Looking at Figure 5-2, you will see that out of the two

Figure 5-1 *Turning a nose cone on a lathe.*

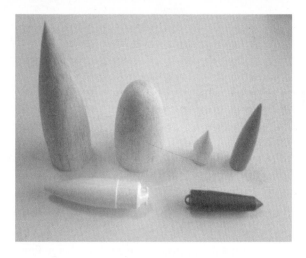

Figure 5-2 *A selection of commercially available nose cones.*

Figure 5-3 *Plastic nose cones often require assembly.*

plastic nose cones at the bottom, the nose cone to the left was manufactured using blow moulding, as a single-piece assembly, whereas the nose cone to the right was made as two separate pieces which need to be glued together. (Figure 5-3).

Always use an appropriate glue for the plastic you are joining. Poly cement works well most of the time. There are a number of ways you can apply the glue without getting messy. Either run a small bead around the shock cord mount, or apply a drip of glue to a piece of paper, and turn the nose cone lip in the glue. Either way, don't go squeezing masses of glue onto the two parts, you will only make a mess! Also, make sure you allow the glue to dry before either:

- attaching the recovery cord – you are likely to pull the mounting away from the nose cone

- pushing the nose cone into the body tube – resist the temptation! It will only get stuck at this point.

Body tubes

The body tube provides the main structural backbone of our model rocket. It is easy to overlook the essential function the body tube performs. In addition to supporting all the components for our model rocket, it also provides a barrier to the elements, and a flame-resistant barrier for the hot exhaust gases emanating from the recovery charge.

The body tubes we use for model rocketry are designed to be flame-retardant and have a lot of strength for very little weight. The tubes are wound from fire-retardant paper in a helical manner – that is to say diagonally. This winding is often clear in a new unpainted body tube. In order to improve the finish of your model rocket as well as lower the overall drag of your model, it is recommended that you apply the tiniest bit of filler between the adjacent windings, to fill in the small ridge between windings.

Body tubes are available in a plethora of different diameters; most of these are fairly standardized using the metric system. If someone refers to a BT-20 body tube, they are referring to a body tube that is 20 mm in diameter.

If you are building a model rocket with a high internal volume (using a large body tube), you might find that the recovery charge generates insufficient pressure to pressurize the container, fill the body tube, and push the nose cone out. In this case, you need to remove some of this volume. A simple way of doing this is to create a bulkhead immediately after the motor mount, with a narrow tube running within the larger tube, stopping clear of the recovery system chamber. An identical bulkhead is positioned here. All joints must be glued to both the internal body tube and the external body tube. This will not be visible from the outside of your model, but what it serves to do is to displace some of the internal volume, removing it from the equation,

Figure 5-4   *An array of different body tubes.*

Figure 5-5   *Right-angled plastic stock. Metal stock may also be used.*

so the volume of recovery gas has to occupy a smaller volume, and hence creates a higher pressure.

If you need to make your own body tube of an unusual size, find a wooden rod, cardboard tube or length of plastic pipe as a former, and wrap the rod with waxed paper. Then take some 1/32 in thick balsa sheet and soak it in water or steam it with a kettle. This will make it more flexible and supple. Wrap a layer of balsa over the waxed paper, and when this is dry repeat with another layer of balsa. Once this step is complete, glue it all together and wrap the assembly with a little silk-span sheeting, commonly used in aircraft modelling, and available from suppliers of aircraft models.

Figure 5-6   *Clamping the body tube in order to mark it for fins or a launch lug.*

Marking lines on body tubes

There has been a nasty rumour in the model rocketry world for some time, that the best way to mark a straight line on a body tube is to "hold the body tube up against right angled edge like a door frame and draw a straight line." Now I don't know about you, but parents soon get mighty frustrated when their beautiful white-gloss door frames become emblazoned with the marks from your latest space exploration endeavour. Far better to keep the peace and invest in a small piece of right-angled stock, not unlike that shown in Figure 5-5. This can be plastic or metal, and will prove to be as useful as a ruler.

To mark a straight line, either hold the tube or lightly clamp it in some way to support it while you draw a horizontal line, as shown in Figure 5-6.

Fins

In order to space the fins effectively, use a fin marking guide. Take a copy of Figure 5-7 and pin it to your bench. The angles are marked out for three-fin, four-fin and five-fin model rockets. Simply center your body tube in the middle of this guide, make a small mark at the base of the tube, and then continue this mark using the method above. Under no circumstances try and guess positioning or judge it by eye – this is only likely to result in disappointment. Ensure when your fins are affixed that they align with the lines that radiate from the center. When your fins are dry and have set in position, simply give them a little fillet between the

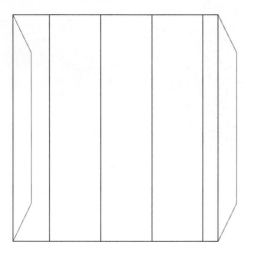

Figure 5-7 *Fin alignment guide.*

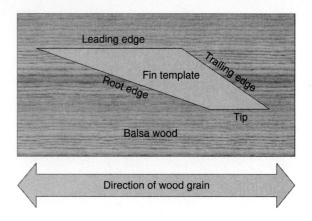

Figure 5-8 *How to orient a fin when cutting from balsa wood.*

body and the fin as shown in Chapter 4: Model Rocket Stability; this will greatly improve the aerodynamics of your model.

You can make fins from a number of different materials. Some entry level model rockets will come with a one-piece moulded fin assembly which comprises the fins and motor mount, and simply push onto a model rocket body tube.

The criteria for a fin material is that it is stiff and rigid. Many materials are good candidates. Artist's mounting board is a good one for quick and easy model rockets. Balsa is better for the skilled modeler because it is possible to put a profile on the fin as well as shaping it to aid its aerodynamic performance.

When cutting fins from balsa wood, it is important to make the best use of the properties of the material, in order to ensure that your model is strong. You need to realize that balsa has a grain, as with all woods, but because the material is so light, it snaps very easily along the grain. In order to get round this, we orient our fin so that the strength is not lost (see Figure 5-8). If you are unfamiliar with fin terminology, look back to Chapter 4: Model Rocket Stability to see what it all means.

If you are using modeler's dope to fill the grain of your fins, it is best to use a flat ox-hair brush of at least 1/2 in (1 cm) width. Load the brush with dope and apply it in quick strokes. Sand the balsa with a fine

400-grade wet and dry paper. Keep repeating until a fine finish is obtained. If you start with a coloured dope, apply a protective layer of clear dope before you finish. If you find the dope on its own is a little thin, you can make a very satisfactory filler from clear modeller's dope, a little dope thinner and talcum powder.

Transitions, bulkheads, and motor spacing rings

If all rockets were dead straight, things would be boring! Quite often, we need a transition between two body tubes of a different diameter. This could be because of visual effect, because we are copying a scale model and we are trying to mimic the design, or because we have a motor that is smaller than our body tube and we need to mount it securely.

We need some sort of transition between the two diameters.

There are many different solutions to the problem, model rocket suppliers will happily sell you all manner of cardboard tubes, rings, and spacers to enable you to reduce (or increase) from one tube diameter to the next; however, making them isn't that difficult!

What you need to remember, if you are intending to cut them yourself, is that the circle that you cut needs

Figure 5-9i *Making motor mounts from MDF circles.*

Figure 5-9iii *A pair of balsa wood transitions.*

Figure 5-9ii *The finished transition – every bit as good as one from the shop.*

Figure 5-10i *Making a slit for the motor hook.*

to fit INSIDE the larger body tube, and the hole you drill in it needs to fit OUTSIDE the smaller tube. Master this and it all becomes easy! Invest in a set of tank cutters, shown in Chapter 3: The Model Rocketeer's Workshop and cut some circles of wood from either thin MDF or plywood that are equal or ever so slightly smaller than the larger body tube diameter. Remember that you can always remove material with a little sandpaper, but it is much harder to replace it. Then, drill a hole through the center of the circle that is slightly larger or the same as the external diameter of the smaller body tube. With a bit of sanding and finishing, you can turn out a nice transition such as that shown in Figure 5-9ii.

Of course, sometimes a flat transition is inappropriate. If the transition is external and subject to airflow, you want the join between the two different tube diameters to be as subtle as possible. Turned balsa wood transitions such as those in Figure 5-9iii are readily available, or using a lathe you can turn them yourself – as with the nose cones!

Motor mounts

The idea of a motor mount is to provide some mechanical support for the motor, stop it moving during either launch or recovery phases, and to provide a means of removing and changing motors easily. At its simplest, a motor mount is a motor hook and a tube (Figure 5-10); increasing in complexity, motor mounts can incorporate some sort of transition element, to allow the motor to be fitted in a larger diameter body tube, or at the height of the hobby, can include mounts for multiple motors with varying diameters.

To construct a simple motor mount, simply take an appropriately sized tube for your motor. Figure 5-10i illustrates making a small slit in the tube for the motor hook, and with the motor hook finally inserted, your mounting will look something like Figure 5-10ii. This could be inserted in a body tube of the correct diameter, or used with a transition piece if it were to be inserted into a larger body tube.

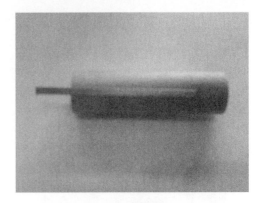

Figure 5-10ii   *The motor hook inserted into the motor tube.*

Project 8: Building Your First Model Rocket

We now move on to the REALLY exciting stuff – building your own model rocket!

You will need

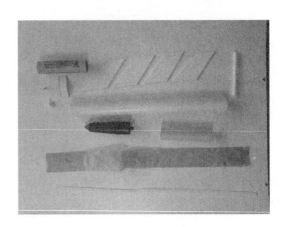

Figure 5-11i   *Components of a basic model rocket.*

- Nose cone
- 12 in (30 cm) length rubber shock cord
- 12 in × 1 in (30 cm × 2.5 cm) streamer fabric/plastic
- Launch lug
- 8 3/4 in (22 cm) length of 18 mm/3/4 in launch tube
- 18 mm model rocket motor (A8-3 recommended for first flight)
- Artist's mounting board

Assemble all the parts shown above.

Using the fin guide introduced earlier in the chapter, mark four points around the circumference of the tube and then using a right-angled piece of plastic, extend those lines along the tube. Now, tape the tube to a table or other study support using a very low-tack adhesive tape such as masking tape. If you made the rocket stand featured in Chapter 3: The Model Rocketeer's Workshop, now is an ideal time to use it!

Once this is complete, glue the motor mount together as described earlier in this chapter. Glue it into the tube.

Figure 5-11ii   *The finished rocket.*

Now assemble the nose cone. This needs to be glued with a little poly cement. Take the supplied shock cord and fabricate a suitable shock cord mount out of a little piece of card. Glue this inside the model and attach the other end to the nose cone. Now tie the shock cord around one end of your streamer.

The streamer should be rolled and inserted into your model, followed by the nose cone. Now glue the all-essential launch lug onto the side and once that is dry you are ready to rock and roll. If you want, you can paint the model and apply a superior finish before launch. Figure 5-11ii shows the finished model.

Project 9: Building a Series-staged Rocket

You will need

- 9 in (20 mm) diameter body tube
- $3\frac{1}{2}$ in (20 mm) diameter body tube
- Balsa sheet
- Model rocket nose cone PC-20A
- 2 foot (60 cm) elastic for shock cord
- 1 × metal eye
- 12 in (30 cm) parachute
- Shroud line string
- Ring reinforcers

Using the fin guide, mark out the positions for the three fins on both the main body tube and the booster section. Extend these lines using a right-angled edge for several inches down the body tube. Work in pencil; it is easier to erase later.

We are going to assemble this model in two parts. Start off with the main rocket. You need to glue the shock cord mount to the nose cone, then tie the shock cord to the mount once it is dry. Then cut out a small piece of card, and mount this onto the main body tube.

Now take your parachute, and cut out the shape, apply ring reinforcers to the corners, and tie the shroud lines. Gather the shroud lines into a bundle and tie them to the nose eyelet, now attach the shock cord to the nose eyelet also. Take the end of the shock cord and glue it inside your model.

Now we need to assemble the booster stage. This is simply a matter of gluing the motor mount into the booster stage body tube and then gluing the fins to the outside of the tube.

You can now paint and finish the model.

Now comes the tricky part, loading the motors! What we want is a motor with a XN-0 designation where X is a letter, N is a number and 0 is the delay and also a standard motor. With a little masking tape, join the two motors together so that the standard motor is on top. Once you have done this, the first motor can be pushed into the main rocket body. Push the motor clip in so that it just breaks the surface of the masking tape. Now slide the booster over the remaining motor.

Figure 5.12 *First stage of series-staged rocket.*

Many rockets are ripe for conversion into series-staged models; what you need to do is add an additional booster stage to the bottom of the model.

This booster stage will have an additional length of body tube as well as an engine mount tube and some fins (Figure 5.12).

Project 10: Build Your Own Parallel-staged Model

If you have worked through the projects so far, you should be fairly fluent in model rocket construction. Here we are going to look at one of the most challenging projects yet – a parallel-boost rocket.

You will need

- 14 in (20 mm) diameter body tube
- 2 × 3 in ½ in (20 mm) diameter body tube
- Balsa sheet
- Three model rocket nose cones PC-20A
- 2 foot (60 cm) elastic for shock cord
- One metal eye
- 12 in (30 cm) parachute
- Shroud line string
- Ring reinforcers

We are not going to go over the details of building the main rocket as this should be fairly simple by now; however, the one thing that needs to be changed is two slots either side of the body tube, an inch above the motor mount. Below this slot, we are going to glue a small filler of balsa wood about $\frac{1}{8}$ in (3 mm) square. This serves to space the parallel boosters from the main body of the model. At the bottom of the fillet, cut a little groove. This serves to locate the pin in the aft mounting of the booster.

The booster construction is very simple. The body tube consists of a motor mount inside a body tube topped out with a nose cone. There are no fins; however, there are two lugs on the same side of the booster, top and bottom. The lugs are constructed from a sandwich of balsa wood. There are two trapezium-shaped pieces either side of the piece of hardware that

engages with the main body tube. At the top of the first is a balsa hook that engages with the slot in the body tube. At the bottom is a pin that holds secure onto the fillet of wood on the main body.

The boosters should fit onto the model so that when they are pushed forward they are securely mounted; however, as they fall back they disengage. A little Blu-tack can be used to steady them for launch. Because the boosters have a high aerodynamic drag when they are expended, they peel away from the main body. The rest of the rocket is self-explanatory.

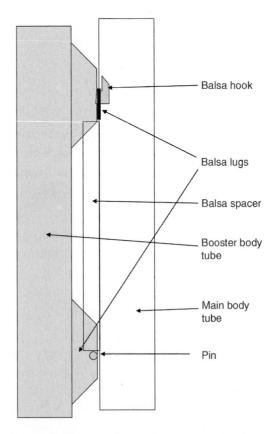

Figure 5-13   *Parallel rocket mounting hardware.*

Balsa hook

Balsa lugs

Balsa spacer

Booster body tube

Main body tube

Pin

Project 11: Build Your Own Boost-Glider

You will need

- Edmonds Aerospace Deltie B Rocket Launched Aircraft kit.

I highly recommend that you start with a kit before attempting to build your own boost gliders from scratch. Once you have mastered the basic principles, you can progress onto more complex self-designed challenges.

The Deltie B is a great kit for you to start off with. The assembly entails construction of a balsa wood glider as shown in Figure 5-14. If you have experience of making balsa wood gliders, you may be able to fabricate something similar yourself. The crucial difference between a boost glider and any other is that the boost glider has a hook on the front which engages with a hook on the booster.

There are a number of boost glider configurations. These are shown in Figure 5-15.

The booster is nothing more than a simple short model rocket without fins. Instead, a balsa wood hook which engages with the glider is mounted to the back of the booster. This is shown in Figure 5-17. Once the motor is expended, the aerodynamic drag causes the two parts to separate. The glider returns to earth gliding gracefully, and the booster comes down on a streamer.

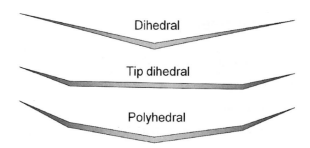

Figure 5-15 *Boost glider configuration.*

Figure 5-16 *Booster motor hook.*

Unidentified flying objects

The existence of unidentified flying objects (UFOs) is well-documented; some say that the sightings of flying saucers are proof of visitors from another planet, others justify the existence of UFOs; by saying that they are spy-planes or the latest models of advanced propulsion under test. Whatever their origin, no one would debate that they are a source of intrigue, and will continue to be so for many years yet.

"Flying saucers" are great fun for the model rocketeer to build; because of their high drag, they accelerate slowly and gracefully into the sky and recovery is generally by tumble recovery – because of their already high drag, they do not need anything further to slow their descent. Nothing quite matches the flight of a UFO; they are exciting to launch and simple to build.

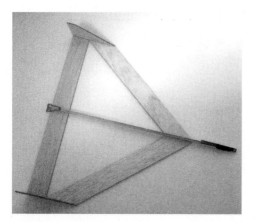

Figure 5-14 *Finished balsa model.*

Project 12: Build your own UFO (I)

You will need

- Polystyrene cup
- Drinking straw
- Motor mount tube

This is a really nice simple little UFO project. To build this cut a circle at the top of the polystyrene cup, glue the motor mount and the straw to the lip. Use an epoxy resin to hold everything firmly in place. Now simply put on the launch pad and fire away. This UFO is so light that it will tumble recover.

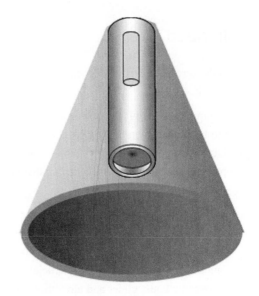

Figure 5-17 *Polystyrene cup UFO.*

Project 13: Build your own UFO (II)

I am indebted to Art Applewhite for allowing me to include his marvelous design in this book.

You will need

- Photocopy of Figure 5-18 in on stiff card
- 1/2A3-4T model rocket motor
- Igniter
- Small motor plug
- White glue

13mm Cinco (Spanish for 5)

1. To making folding easier, crease all the dashed lines with a ball point pin and a straight edge.
2. Cut out the Motor Mount on the solid lines, fold all of the dashed lines
3. Fold and glue Tab 1A to Tab 1B with the printed sides together.
4. Fold Tabs 1A&B over to 1C and glue.
5 Repeat steps 3 & 4 for Tab 2 through 5.
6. Form the Motor Mount into a 5 sided, hollow tube and glue together using Tab 6.
7, Cut out the main body of the Cinco on the solid lines, fold on the dashed lines.
Form the Top into a five sided "cone" and glue.
8. Form the Bottom into a five sided "cone" with the edges folded up and glue.
9. Spread glue on the tab between the Top and the Bottom and press it flat.
10. Spread glue on the 4 remaining tabs of the Bottom and press them against the the Top.
11. Slide the Motor Mount up through the Bottom and continue until it reaches the Top and just a little (1/32") above it.
12. Put a fillet of glue around the Motor Mount where it joins the Top and Bottom.

Recommended motors: 1/4 A3-3T, 1/2A3-2T,A3-4T, A10-3T, A10-PT

Art Applewhite Rockets © 2005
www.artapplewhite.com

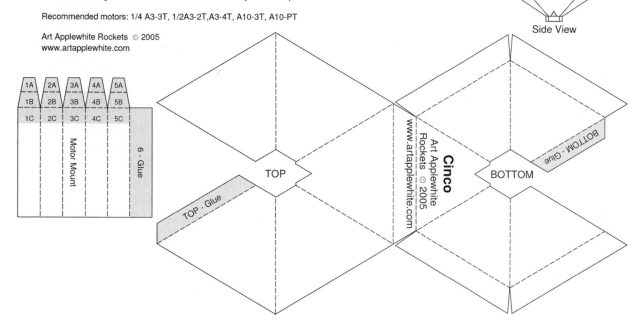

Figure 5-18   *Art Applewhite UFO design.*

You will need

- PVA glue
- Sandpaper

Tools

- Scalpel

Tech Spec

6 in (150 mm) diameter UFO

2.75 in (70 mm) height

0.8 oz (21 g) weight

This UFO is slightly more complicated, and requires some foamboard

First of all, you need to cut out the component parts from those supplied in the kit. Use a sharp knife and take your time – try and ensure that you stick to the lines in order to get the best possible result.

The two cones cut from the fluorescent card need to be assembled. In order to do this, you will need to use the top seam and bottom seam pieces, which should be adhered to the bottom of the cones. Once the cones are assembled, move to the section of foamboard. Cut out the centre using a sharp scalpel; be very careful as the positioning and quality of the motor mount cut is essential – this will be the part that supports the motor and ensures the safe flight of the rocket. The foamboard centre of the UFO requires a little bit of additional work – you will need to dig the scalpel into the board, and cut at an angle towards the outside of the shape. Be careful when doing this, there is a lot of potential to cut yourself here.

Now glue the cones either side of the foamboard using the motor tube as a guide to align everything. Ensure that you lineup the holes for the launch rod.

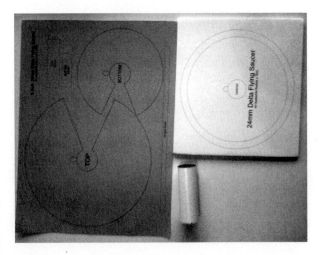

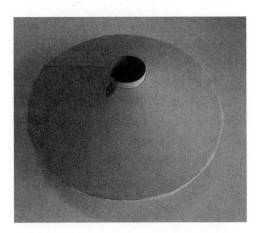

Figure 5-19 *Foamboard UFO design.*

Project 15: Build Your Own Payload-carrying Rocket

You will need

- Nose cone (length 70 mm, diameter 45 mm) (see template, Figure 5-20)
- 125 mm polycarbonate body tube
- 45–40 mm transition length 15 mm (see template, Figure 5-20)
- 540 mm body tube
- Recovery system of choice
- D motor mount rings
- Balsa for fins (see template, Figure 5-20)

The payload-carrying rocket is of a very simple construction. Everything is the same as the simple rocket, with the exception that everything is bigger! You will notice that the payload is carried at the front in a polycarbonate tube. We do not use quantum tubing because some of the projects require the payload to be able to "see" the outside world.

Most of this assembly is press-fit with the exception of the motor mount, which must be glued.

When selecting your recovery system, you need to size your parachute so that it will cushion the landing of your payload.

This rocket is incredibly stable in flight – it is designed to be overstable; however, this can be compensated for by adding weight where necessary. The rocket launches very slowly using a D-12, you might want to try it with something a bit meatier for a heavier payload.

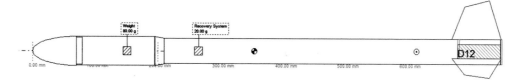

Figure 5-20 *Payloader rocket dimensions.*

Chapter 6

Recovery Systems

A wise man once said "What goes up must come down." There is a huge body of scientific knowledge and experiential data to support this simple fact. The same must be said for model rocketry. Your rocket has launched successfully and flown high into the air. At this point comes a simple choice:

a) Your rocket can descend gracefully to earth, allowing recovery of all components intact with paintwork unscratched and investment undamaged – ready to fly another day.

b) Your rocket can float off gracefully into the distance, like dandelion seeds on the wind, and snag on a tree, pylon, or migrating goose somewhere the other side of the state border.

c) Your rocket can hurtle to earth at a million miles an hour (artistic exaggeration) and firmly embed itself in the upper mantle of the earth.

If you fancy flying without a recovery strategy, think option c) If however, you are looking at a) and b) then think of a recovery system – a) is the result of a well-designed recovery system, b) is the result of no calculation, or an over-zealous hobbyist.

Recovery system basics

The recovery system is the series of components that will bring your rocket to ground safely. The two most common recovery systems – streamers and parachute – maintain all of the components, being connected by a shock cord.

See Figure 6-1 for an outline of the various components. You will see that the nose cone is separated from the main body of the rocket. The force of the gas generated by the ejection charge causes the rocket to separate. This is dramatic and on low-powered rockets you can clearly see the parachute being pushed out of the body tube. This is illustrated dramatically in Figure 6-2.

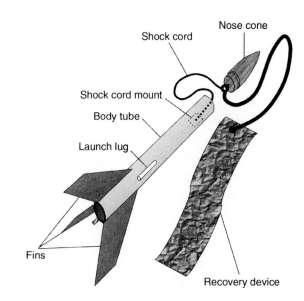

Figure 6-1 *Outline of the recovery system.*

Figure 6-2 *The ejection charge deploys and the parachute is ejected.*

In order that the two components stay together, the nose is attached to the rocket body by a shock cord. This is generally made from either rubber or elastic in order to absorb the force generated by the rocket separating. The shock cord is further attached to the recovery device where fitted.

Recovery system types

Nose blow recovery

Nose blow recovery is one of the simplest recovery systems. Essentially, the rocket's aerodynamic form is spoiled by the nose detaching from the main body in flight. The nose remains tied to the rocket body by a shock cord. Nose blow is not the best recovery method out there – while it slows the rocket considerably, it does not always slow it enough to guarantee safe recovery of the model.

Featherweight recovery

Featherweight recovery is used for small, light, model rockets. In this type of recovery system, the motor is ejected from the main body of the model rocket. The remainder of the rocket, falls at a slow speed because the model weighs next to nothing, but its surface area is high. This results in a high surface-area-to-weight ratio.

You won't see featherweight recovery at organized contests because most major rocket clubs will not permit free-fall motor casings.

The UFO models in Chapter 5 are good examples of featherweight recovery. See Figure 6-3.

Figure 6-3   *UFO featherweight recovery.*

Tumble recovery

Tumble recovery is where the engine is either ejected or thrust rearward by the ejection charge. This upsets the balance of the model, and causes it to tumble to ground.

Streamer recovery

Streamer recovery involves deploying a long streamer, usually made of plastic, in order to slow the rocket down. Generally, streamers function well with a length-to-width ratio of 10:1, according to research performed at the Massachusetts Institute of Technology with Trip Barber.

How you roll your streamer is up to you; different people have different techniques. Some coil their streamers into a spiral, while others wind them back and forth in a zig-zag pattern.

Figure 6-4i   *Streamer attachment to shock cord.*

Figure 6-4ii   *Streamer rolled and inserted into nose cone.*

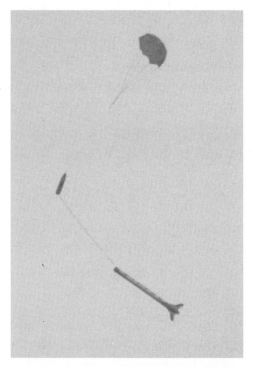

Figure 6-5 *Parachute models descend gracefully from the sky.*

Once the streamer is rolled, it is pushed gently into the body tube. It should be free to slide out without any restriction or snagging, so that it can be deployed easily when the ejection charge fires.

As well as plastic, streamers can be cut from crepe paper, tracing paper, or Mylar film.

The streamer is attached to the shock cord near the nose cone, as illustrated in Figure 6-4i. It is then rolled or pleated and inserted into the nose cone (Figure 6-4ii).

Parachute (parasheet) recovery

Most rockets use a parasheet – this is just an approximation of a parachute using a flat sheet generally; this would still be referred to as a parachute. Parachutes are available cheaply in premarked patterns, although you can make your own quite easily and economically. Better parachutes are made of panels and form a hemisphere when they unfold. Parachute models can look really graceful as they descend from the sky.

The larger the diameter of your parachute, the more drag it will create, hence the slower your model will travel. There is a real trade-off here – slow your model down so it is guaranteed to land safely and you run the risk of it floating away over the trees! Aim for a fast descent and you run the risk of the model breaking when it lands. Table 6-1 is a quick and easy ready-reckoner.

For cheap plastic parachutes, attaching the shroud lines, that is to say the fine strings that make the parachute assume its shape, is a matter of using some

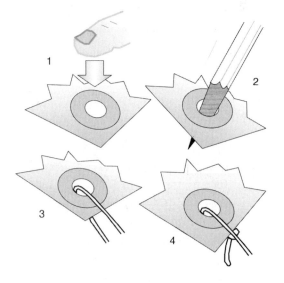

Figure 6-6 *Attaching shroud lines to a paper parachute.*

Table 6-1
Parachute Diameter Ready-Reckoner

Weight		Parachute Diameter				
(Grams)	(Ounces)	1 m/s	2 m/s	3 m/s	4 m/s	5 m/s
10	0.4	0.37	0.19	0.12	0.02	0.01
20	0.7	0.53	0.26	0.18	0.02	0.02
30	1.1	0.64	0.32	0.21	0.03	0.02
40	1.4	0.74	0.37	0.25	0.03	0.02
50	1.8	0.83	0.42	0.28	0.04	0.02
60	2.1	0.91	0.46	0.30	0.04	0.02
70	2.5	0.99	0.49	0.33	0.05	0.02
80	2.8	1.05	0.53	0.35	0.05	0.02
90	3.2	1.12	0.56	0.37	0.05	0.02
100	3.5	1.18	0.59	0.39	0.06	0.03
110	3.9	1.23	0.62	0.41	0.06	0.03
120	4.2	1.29	0.64	0.43	0.06	0.03
130	4.6	1.34	0.67	0.45	0.06	0.03
140	4.9	1.39	0.70	0.46	0.07	0.03
150	5.3	1.44	0.72	0.48	0.07	0.03
160	5.6	1.49	0.74	0.50	0.07	0.03
170	6.0	1.54	0.77	0.51	0.07	0.03
180	6.3	1.58	0.79	0.53	0.07	0.03
190	6.7	1.62	0.81	0.54	0.08	0.03
200	7.1	1.67	0.83	0.56	0.08	0.03
210	7.4	1.71	0.85	0.57	0.08	0.03
220	7.8	1.75	0.87	0.58	0.08	0.03
230	8.1	1.79	0.89	0.60	0.08	0.03
240	8.5	1.82	0.91	0.61	0.09	0.03
250	8.8	1.86	0.93	0.62	0.09	0.03
300	10.6	2.04	1.02	0.68	0.10	0.03
350	12.3	2.20	1.10	0.73	0.10	0.03
400	14.1	2.35	1.18	0.78	0.11	0.04
450	15.9	2.50	1.25	0.83	0.12	0.04
500	17.6	2.63	1.32	0.88	0.12	0.04
550	19.4	2.76	1.38	0.92	0.13	0.04
600	21.2	2.88	1.44	0.96	0.14	0.04
650	22.9	3.00	1.50	1.00	0.14	0.04
700	24.7	3.12	1.56	1.04	0.15	0.04
750	26.5	3.22	1.61	1.07	0.15	0.04
800	28.2	3.33	1.67	1.11	0.16	0.04
850	30.0	3.43	1.72	1.14	0.16	0.04
900	31.7	3.53	1.77	1.18	0.17	0.04
950	33.5	3.63	1.81	1.21	0.17	0.04
1000	35.3	3.72	1.86	1.24	0.17	0.05

paper ring reinforcements, like the sort used to strengthen documents where ring-binder perforations are broken through. Simply apply a paper reinforcement to the position where you want to attach a shroud line. Pierce with a pencil, and then anchor the fine shroud line. Simple really. This is shown in Figure 6-6.

Helicopter recovery (autorotation)

Helicopter recovery models are absolutely fascinating to watch and represent a VERY worthwhile investment of your time and effort. A helicopter recovery rocket takes off like a regular model rocket. When it reaches its apogee, the recovery system deploys. The recovery system takes the form of a number of blades which spring out from the model rocket. The weight of the model makes it fall to earth, but the air rushing over the helicopter blades causes the model to spin as it descends.

If you are interested in building a helicopter recovery model, the best place to start is with a kit. They can be quite hard to get right first time if you are building from scratch.

Glide recovery

Glide recovery models soar into the sky as a rocket, and then glide back to earth as a plane. A boost glider, is effectively a rocket which acts as a booster, and a glider which detaches and glides back to earth.

Shock cord mounts

There are a number of ways of mounting shock cords. Two of the most common are:

Folded paper method

The shock cord is folded within a small piece of paper which is then glued to the inside of the model. See Figure 6-7.

Figure 6-7 *Folded paper shock cord mounting.*

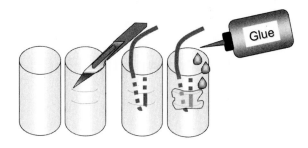

Figure 6-8 *Through-tube method of mounting shock cords.*

Through-tube method

The shock cord is passed through two small slits in the body tube, knotted, and then glued over.

Shock cord materials

The shock cord ties the nose, body, and recovery system of your rocket together. It needs to be able to withstand the stress of being ejected from the tube, and to have the tensile strength not to snap when the body and the nose move in different directions.

For smaller rockets, plain old vanilla elastic or rubber will do just fine. Estes have recently made the move to rubber recovery cords as they believe that they perform better; however, many people still use elastic cords with a fair degree of success.

For really exotic rockets, we can even push the boat out and use materials such as Kevlar. Such exotic materials aren't disproportionately expensive; Kevlar cord (Figure 6-9i) has a tensile strength of hundreds of pounds, and so for some high-powered rockets can be a worthwhile investment

For really heavyweight models, Kevlar tape is even stronger than Kevlar cord (see Figure 6-9ii).

Figure 6-9 i   *Kevlar shock cord.*

Figure 6-9 ii   *Kevlar shock tape.*

Recovery wadding

Remember those hot ejection gases? Well, they are almost as hot coming from the recovery system as they are from the exhaust! To protect our delicate little recovery systems from this onslaught of hot gases we use recovery wadding. Recovery wadding puts a flame-resistant barrier between the hot exhaust gases and the recovery system. It is just like a tissue that has been treated with flameproof chemicals. Look at Figure 6-10, this is how recovery wadding is supplied.

Project 16: Build a Night-Tracer Rocket

You will need

- Tired model rocket in need of new life
- Electroluminescent panel (see suppliers' index)
- Small inverter
- Battery clip
- 9V battery

or

- EL inverter from Project 17 – Build Your Own EL Inverter

Tools

- Soldering iron
- Scalpel

Figure 6-10   *A packet of recovery wadding.*

Figure 6-11   *Night-tracer rocket.*

There are compelling reasons for launching rockets in the dark. Distress flares have been used for emergency purposes for many years to alert rescue teams to the presence of people in danger.

You might want to build this project "for emergency use only," or you might have a compelling reason for wanting to launch a rocket in low light – maybe you live near the North Pole and don't receive much daylight; maybe you want to photograph your rocket against a glorious sunset, or backlit by the moon and stars; maybe you find the wind still at night; or perhaps you simply enjoy the cool night air.

Whatever your motivation, you can bet your bottom dollar that if you fire your cardboard contraption into the dark abyss, the chances of it being found are slimmer than a model-turned-actress after a week of cabbage soup.

What you really need is a night-tracer rocket as illustrated in Figure 6-11.

Luckily help is on hand with this project, which harnesses the property of electroluminescence to great effect.

Tell me more about EL...

Electroluminescent (EL) panels, also known as light-emitting capacitors, produce light when phosphors are excited by electrons, producing photons – the light we see in the process.

The reasons that these panels make good night tracers is because there are no fragile mechanical parts to break – unlike an incandescent lamp with a fine filament – furthermore, they are flexible, which allows us to wrap them around our rockets. The EL panels are *sooo* thin that they add negligible weight to your rocket, the main weight being the power supply.

Obviously as your rocket takes off there is a whoosh of smoke, and suddenly your creation disappears up into the clouds. The beauty of EL panels is that, as they have such a large surface area, they are easier to see in adverse conditions than a simple point source of light.

How does it work?

When EL phosphors are placed in the field created by an alternating current, they will emit some light. As the voltage is increased, so is the brightness, albeit at the expense of the lifetime of the phosphor. Turning the phosphors on and off will not damage them; however, they do degrade with continual use.

There are two types of phosphor, encapsulated and unencapsulated. As you don't know where your rocket is going to land, or how the weather is going to change, I would strongly advise you to plump for the encapsulated polymers – these are not affected by moisture to the same degree as unencapsulated EL panels.

How to make your night-tracer rocket

Electroluminescent panels require a fairly high AC voltage in order to work effectively. We can produce a high-voltage AC discharge from the EL inverter described below. Once you have constructed the inverter, making a night-tracer rocket is a matter of covering the outside of your rocket with a suitably

stuck-down EL panel or wire, and connecting it up to the inverter carried internally as a payload for your rocket. Remember to consider how the weight of the payload will affect the weight distributions.

Taking it further

With your night tracer set up, there is plenty of room for additional experimentation; for example, using a camera set up on a tripod, some distance away from the rocket, it is possible to calculate the speed of descent of the parachute by using a timed exposure, and calculating how far the object travelled in that exposure. By using a small aperture and your camera's "bulb" setting, you can photograph the descent of your rocket, and note how the wind affects the descent of the parachute. Using a wide-angle lens to take in a lot of the scene, you can photograph the ascent of your rocket as a streak through the night sky.

Project 17: Build Your Own EL Inverter

You will need

- MC 33441 EL driver chip
- 130 K resistor
- 0.1 mF capacitor
- 27 nF (100 V) capacitor
- 1 mHz coil
- 2 AA battery holder
- 2 AA batteries
- EL panel/wire

Tools

- Soldering iron

It is possible to build you own simple and effective EL inverter using the MC 33441 driver chip; this requires a minimum of external support components in order to make an effective EL inverter.

This circuit runs on a pair of AA batteries, the schematic is shown in Figure 6-12. Be careful as this project generates a high voltage that has the potential to shock. Try to mount it in some sort of plastic enclosure to protect yourself. Tic-Tac boxes are cheap, readily available, and fit in most larger body tubes.

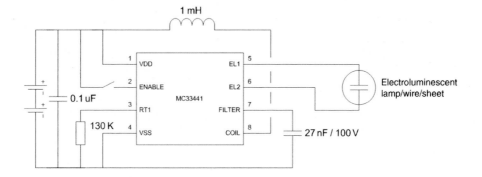

Figure 6-12 *Schematic of EL inverter circuit.*

Project 18: Creating a Strobe Beacon Rocket

You will need

- Kodak Flash disposable camera
- Teccor K1300E70 sidac
- Transparent body tube or nose cone for mounting

Tools

- Small flat-blade screwdriver

A strobe is a really useful tool to enable you to find and recover your model rocket, as well as making an interesting electronic payload. By flashing at regular intervals, your model rocket can be seen more clearly during launch, descent, and eventual recovery.

The problem with most strobe schematics and kits on the market is that they either require line voltage, or at very best 12 V to operate. This is a bit of a pain for a model rocketeer as it means a number of batteries – let's face it, running a mains cord to your rocket isn't smart, nor is it going to get very far!

This project looks at modifying a disposable camera, cannibalizing its flash unit, and turning it into a permanent strobe.

This means that you can work with a ready-made unit with a little modification, saving on component cost.

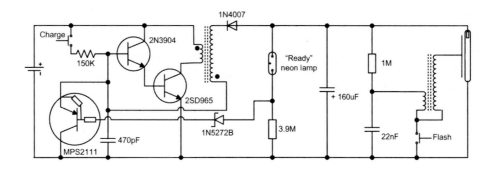

Figure 6.13 *Kodak MAX Flash schematic.*

These things are made wholesale and very cheaply; you can pick up a cheap disposable camera with flash for around $5 so it hardly makes shopping for components worth doing.

For the hardcore electronics enthusiasts, I have also included the schematic of the flash unit on my Kodak MAX disposable camera as a base for experimentation with your own circuits.

Caution

Do not assume that because you have taken the battery out, the circuit is safe to work on. This project contains a capacitor that will have high voltages stored in it when you prise apart the camera. Even with the battery removed, this circuit still has the potential to shock – a lot!

They say that it takes 10 J to kill someone with relative certainty. Even so, it is well worth avoiding shocks if at all possible, as smaller amounts of energy still have the potential to do damage. At any one time, this flash unit will contain about half of that amount, still enough to give a real belt. The capacitor inside the flash unit WILL be charged to several hundred volts. It you touch it it will REALLY REALLY REALLY hurt. (Believe me, I know). Wear thick insulating gloves. Do not try shocking yourself for fun, nor shock a friend or relative. It really isn't at all funny.

Project instructions

These instructions were designed for a Kodak Flash disposable camera, although they could just as easily work with other brands of a similar design. The reason that I like the Kodak Flash disposable cameras is that after a little investigation with several different brands, I found the Kodak far easier to work on by virtue of the fact that the components are all "pin through hole" rather than surface mounted (as is found notably on Fuji brand cameras).

First of all, don't waste the camera, take it to a party, shoot some photos until the film has been used up. Rather than in a conventional camera where the film is unrolled into the camera with each successive shot, then wound back into the camera at the end, in a disposable camera, the film is already unrolled inside the light-tight camera, and each time you "wind on" you are winding the film back inside the cartridge.

Once the film is used up, you can remove the cardboard outer covering of the camera, exposing the black plastic body. You will see a number of black plastic tabs all over the body of the camera. With a little effort with a small flat-blade screwdriver, you can pry these tabs away from the camera body one by one. You should be able to remove the film with relative ease.

Now remove the battery from the camera. In the Kodak Flash models this is a single AA battery located behind a flap on the left hand side of the camera. You will need to "persuade" the tabs with a small screwdriver. On some other models, I have seen the battery located in the base of the camera.

You now need to remove the small circuit board from the front of the camera. As noted in the box YOU MUST wear thick insulating gloves as you will definitely get a nasty shock otherwise.

The first step is to discharge the capacitor. This can be done using an old screwdriver. You need to locate the large electrolytic capacitor, and look for the joints where the capacitor is soldered to the circuit board. Touch the screwdriver to these connections several times; there will most likely be a spark or pop, but do not worry.

Next you need to disassemble the camera; you want to retain the circuit board and flash unit; we will be rehousing the flash module, but the rest of the plastic is garbage.

We are going to use a pair of sidacs wired in series; the sidacs present an open circuit until a voltage threshold is reached. Once this threshold is reached, the Sidacs switch into a conducting mode – this presents a short circuit.

Note for scratchbuilders

The MPS2111 is sometimes known as a "digital transistor." It is essentially a PNP transistor with additional resistors on its base-emitter. In the event of not being able to obtain the correct part, you may have luck with a boring old 2N3906 and a couple of resistors.

Taking recovery systems further...

In my book *50 Awesome Auto Projects for the Evil Genius*, there is a circuit for a vehicle homing device, that could just as easily be used to locate model rockets when they are on the ground – how's that for a recovery device?!

Further reading

Harper GDJ. *50 Awesome Auto Projects for the Evil Genius*. McGraw-Hill, 2006.

Chapter 7

Launching Model Rockets

When launching model rockets, safety is imperative. In order to perform a safe and successful launch, you need to have a couple of simple items of ground support equipment in order to ensure that your rocket takes off and continues flying safely.

The two main items of ground support equipment required are a launch controller and a launch pad. The launch controller's job is to deliver a suitable amount of electricity to the igniter in the bottom of the rocket engine, to ensure that it lights correctly. We never use fuses or any other means for igniting rocket motors – it simply isn't safe. You never see NASA light the blue touch paper and run like billy-o, so there is no need for model rocket enthusiasts to either!

We are going to start by looking at the part closest to the rocket, the model rocket igniter, then we will take a look at the launch controllers that we can use in order to fire the igniter, and finally, we will look at a few simple designs of launch pad used to guide our rockets straight into the air.

Two launch controllers are presented in this chapter. The first is the simple launch controller; the design is tried and tested and very popular. The second, advanced, launch controller provides more power, and should be used for models where multiple motors are used, or where parallel staging is used and more current is required to fire more igniters.

Both designs feature a "safety key" – this is a key that must be inserted in order for any power to flow. It is there to protect you, and designed so that the launch controller cannot be inadvertently switched to fire.

Model rocket igniters

These little beasties are used to fire rockets safely and consistently; they convert the electrical power from the launch controller into heat energy; this heat energy is used to trigger a chemical reaction in the pyrotechnic lump of material on the igniter.

Simple igniters, such as the Estes Solar Igniter, only require a minimum of 6 V at 0.5 A to fire. They are reliable, and manufactured in order to ensure that they fire consistently and accurately.

Let's take a look at igniter construction, in order to understand how the launch system works. Take a peek at Figure 7-1.

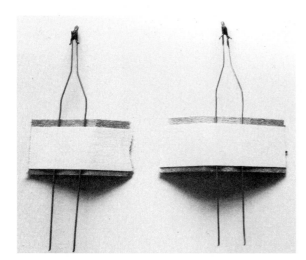

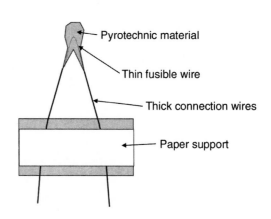

Figure 7-1 *Rocket igniter construction.*

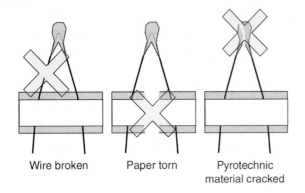

Wire broken   Paper torn   Pyrotechnic material cracked

Figure 7-2 *Common igniter faults.*

You can see that the two fine leads are supported by a paper tape support. This helps to eliminate undue flexing of the connection wires. You should try to bend these as little as possible. While these are fairly study and are unlikely to break, you stand a high chance of damaging the fine wire that triggers the pyrotechnic

material that launches your model rocket; this can be seen at the top of Figure 7-1.

Some common igniter faults can be seen in Figure 7–2. If you encounter an igniter in any of these conditions, you should not use it under any circumstances. You should NEVER attempt to repair an igniter.

Looking from left to right, we can see the first igniter has a broken wire – this will prevent the flow of electricity and the igniter will not trigger. The support paper of the next igniter has torn. While this may seem trivial, flexing of the connection leads may have caused the fine fusible wire of the pyrotechnic material to be broken; as a result, the triggering of your model rocket could be sporadic and unpredictable. The pyrotechnic material on the final igniter has broken.

Igniters should only be triggered when they are correctly inserted inside a model rocket engine. To undertand what happens when a model rocket igniter is triggered, look at Figures 7-3.

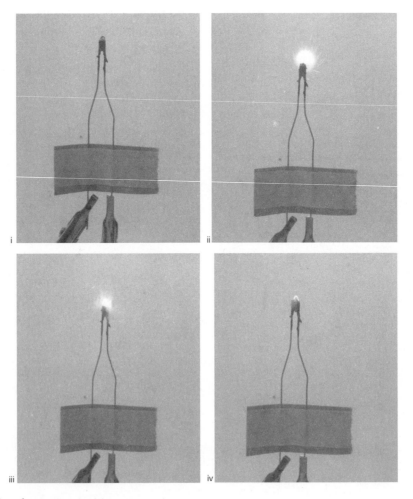

Figure 7-3 *Model rocket igniter ignition.*

Figure 7-3i is taken just as the button is pressed; the fine wire begins to heat up; it is noticeable that it has melted through the tip of the material, and when it reaches a sufficient temperature it ignites the pyrotechnic material which flares. The pyrotechnic material ignites and flares up igniting the model rocket motor. Once the pyrotechnic material is spent, you will be left with either a glowing hot wire, or, more likely, a broken circuit. Because the igniter photographed was not inside a rocket engine at the time of ignition, it has managed to survive intact; this is quite unusual, as normally the intense heat generated by the ignition of the model rocket motor is enough to destroy the fine filament of wire that heats when electricity is passed through the igniter.

Project 19: Simple Launch Controller

You will need

- 12 V battery
- Momentary key switch
- Push-to-make momentary switch
- 12 V LES Bulb (around 1 W)
- LES bulb holder (bezel and enclosure optional)
- Two unscrewable cable connections
- Length of #18 AWG wire
- Crocodile clips
- Project enclosure

Hint: remember it is always easy to cut a wire shorter, however, adding length is more of a pain(!)

Tools

- Drill bit
- Abrafile blade
- Soldering iron

We are going to make a dead simple model rocket launch controller. This project is simplicity itself and you should have very little trouble understanding the circuit's mode of operation. If you refer to Figure 7-5 you can see that we have a circuit that includes a 12 V battery; if we follow this round in a clockwise fashion, we see the safety switch, which is preferably some sort of key switch to prevent interference. The switch only need be a single pole, single throw variety, and preferably is only "momentarily" on when the key switch is turned. You will often find this sort of key switch used in shop front rollers that are actuated automatically when the key is inserted and turned to lower the roller; the roller stops if the key is released. The key switch allows us to "lock off" the circuit in order to connect the model rocket motor safely without fear of inadvertent launch. Remember, it is your eyes and your hands that stand to be damaged if a model rocket goes off inadvertently, so if possible I would always disconnect the battery if it is accessible to ensure there can be no mistake. From the safety switch we move to the launch switch, which is connected in parallel with a small signal bulb. This signal bulb glows when the safety switch is turned and there is a complete circuit with the igniter. It passes enough current to check circuit continuity, without being enough to trigger

Figure 7-4 *Assembled simple launch controller.*

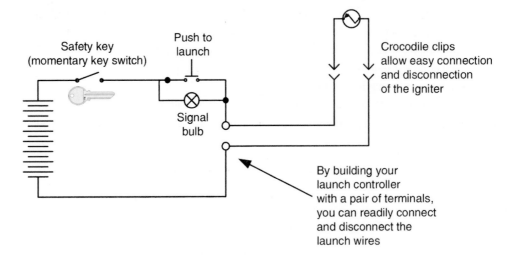

Figure 7-5 *Simple launch controller schematic.*

the rocket. If we press the launch switch, we are effectively bypassing this signal bulb and its resistance. This causes full current to flow to the igniter and launch the rocket. I would strongly suggest that you use a heavy gauge of wire for connecting to the igniter in order that you provide a path of least resistance. In addition to the bulb, I also like an audible warning. This is optional; I have wired a small buzzer inside the case,

which can be used in addition to or substituted for the signal lamp.

There are a variety of cheap plastic housings available at most electronics stores to house your project. Holes are easy to cut and irregular-shaped openings simply require a little chain drilling and finishing off with an abrafile.

Project 20: Advanced Launch Controller

You will need

- Components for the simple launch controller
- 12 V relay
- 2 × banana sockets

Tools

- Abrafile blade
- Soldering iron

The advanced launch controller allows for a larger power output than for the simple launch controller. The

Figure 7-6 *Assembled advanced launch controller.*

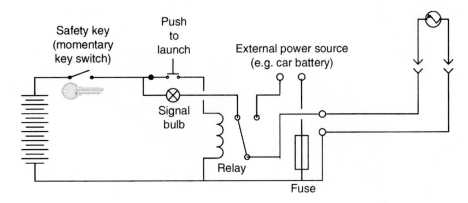

Figure 7-7 *Advanced launch controller schematic.*

technique is similar to that of the simple launch controller, with the addition of a relay to allow the rocketeer to provide an external power source, and also a couple of sockets to allow for connection of this power source.

The finished article looks like that shown in Figure 7-6. The only variation from the simple launch controller is the two additional holes for the external power source. See Figure 7-7 for a schematic diagram.

The relay allows us to use higher-powered batteries. This is especially useful when we want to launch large model rockets. You can make a lead up that allows you to connect the launch controller to your car cigarette lighter socket; this will provide masses of current, surely enough for any model rocket; however, it does necessitate having your car on hand. Failing that, simply carry a lead acid or gel battery into the field with you.

Project 21: Arming Your Rocket Motor for Launch

You will need

- Model rocket
- Model rocket motor
- Igniter
- Motor plug

Model rocket motors are ALWAYS launched using igniters; there simply is no other way of doing things.

Igniters have proved themselves to be consistently safe and as such they have earned themselves a reputation as being THE way to launch model rockets, certainly the only way endorsed by all the national safety codes.

Loading the igniter is as easy as 1–2–3; see Figure 7-8. The motor is shown removed from the rocket for clarity.

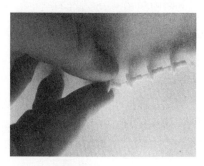
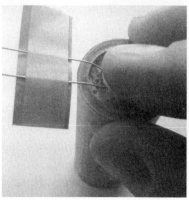

Figure 7-8 1) *Insert the igniter into the throat of the model rocket motor.*
2) *Now take one of the small plastic plugs supplied with your motor.*
3) *Push the plug into the throat as far as it will go and you are ready to rock(et) and roll!*

Project 22: Build a "Rod" Launcher

You will need

- Sheet of plywood 2 feet × 2 feet (60 × 60 cm)
- Stainless steel dinner plate
- 3 mm steel rod
- Solid steel bar

This launcher is really easy to build and provides the basis for a modular launch system that will be extended in the next couple of projects. The base of this launch system is a large plywood disc which sits on the ground, and a large stainless steel plate.

This is purposefully larger than most model rocket launchers, a 2-foot circular plate is easy enough to carry to the field, and because it protects the grass or whatever (flammable) surface your rocket happens to be resting on from burning fragments of igniter, there is no need to carry an additional sheet or tarpaulin with you.

First of all, take a sheet of plywood, and draw a circle as shown in Figure 7-9. An easy way to do this is to fix a screw into the centre of where you require your circle, then tie a pencil to a piece of string, and set the length of string between your pen and the screw to be equal to your desired radius.

Once this has been done, you can cut the shape out with a jigsaw. You will be left with a little scrap. I strongly suggest that you cut out a number of plywood

Figure 7-9   *Drawing a circle using string and a pencil.*

Figure 7-10   *The finished base complete with feet.*

Figure 7-11   *Painting the base.*

Figure 7-12   *The stainless steel blast deflector.*

circles. These can be screwed to the bottom of the plywood disk to raise it slightly from the ground. As our modular system will be bolted to this plywood base, it gives us the clearance for a nut underneath the rocket launcher.

You might also like to drill a few holes around the perimeter of the plywood base. These serve to allow tent pegs to be used to anchor the base to the ground for the two larger launch configurations.

Tie the base to a washing line using one of the tent peg holes and spray it with some black auto paint (see Figure 7-11). This will not only make the base look attractive, but will also protect it and ensure a long life.

We will be using stainless steel as our blast deflector; a cheap way of obtaining a nicely formed stainless steel

blast deflector is to go to the kitchen section of a homewares store, and procure something like the stainless steel plate shown in Figure 7-12. A hole is drilled through the centre of the plate and the centre of the launch pad beneath. For the rod and rail launchers, this is secured with a bolt; however, we need a longer bolt for the rod launcher configurations, as the bolt also needs to pass through a small block of wood. This block of wood, shown in Figure 7-13, features holes to accept our launch rod. The holes are drilled at the vertical and a number of other angles. The rod is pushed into place at the desired angle, and that's about it!

Note: because the rod is quite sharp and could easily impale someone, it is a good idea to use a rod protector

Figure 7-13 *The launch rod mount.*

Figure 7-14 *Rod protector.*

Figure 7-15 *Estes rod launcher.*

(this serves to protect you more than the rod). You can buy one with a length of bright streamer fabric attached, such as that shown in Figure 7-14, or you can make one out of a tennis ball pierced with a hole, or an old film can.

If all this sounds like a little too much hard work, a simple rod launcher comes with many model rocketry starter kits, or can be bought for about $20. The commercially available article manufactured by Estes is shown in Figure 7-15.

Project 23: Build a "Gantry" Launcher

You will need:

- Threaded rod
- Electrical conduit 25 mm diameter (three or four lengths of 2 m or more)

Tools

- Epoxy resin adhesive

A gantry launcher is one of the more sophisticated ways to launch model rockets. The design has the advantage that the launch lug is removed from the model rocket, reducing an element of drag. The trade-off for this is that the launcher must be a lot more complex.

The design set out below is simple and successful, and can easily be adapted for four-fin and three-fin rockets.

The design is as follows. Depending on whether the design is for a three-or four-fin launcher, the rocket will be supported by three or four lengths of plastic conduit. The plastic conduit is held in place by a couple of lengths of threaded rod, which allow

Figure 7-16   *"Stripy" threaded rod.*

Figure 7-17   *Small nut glued onto the end of a threaded rod.*

Figure 7-18   *Hole and slot cut into electrical conduit.*

adjustments to accommodate different sizes of rocket. The threaded rod is in turn supported by a metal frame at the top, to ensure that the rods stay parallel, and a frame at the bottom, which incorporates a blast deflector.

The first stage of the project is to cut six or eight (depending on what launcher configuration you are building) threaded rods of the same length. I suggest 10 in (25 cm) as a good size to give you plenty of rod to work with and to make the assembly easy and manageable. Once you have cut these lengths of rod, you need to finish their ends with a file to allow a nut to be easily screwed on from each end.

Next, take the rods, a ruler, and a permanent marker, and make regular marks every half inch or centimeter, depending on whether you work in imperial or metric. These marks are arbitrary, but enable you to align and calibrate the rods more easily when adjusting the gantry launcher to accommodate different-sized models.

Next, you need to color alternate bands on these rods so you are left with a rod that resembles that shown in Figure 7-16.

The next step is to glue a nut onto the end of each rod using an epoxy resin adhesive. The nut should be locked right on the end of the rod and should not turn when the glue is set. The ends should then look something like Figure 7-17. These ends will engage with slots that are cut into our launch poles. It is these poles that will guide the model rocket safely into the air.

We now turn to our lengths of electrical conduit. Measure about one-and-a-half foot (50 cm) from each end, and make a mark on each length of conduit at that point.

We are going to drill a hole to enable us to slot the tubes over our threaded rods, which will in turn be supported by brackets. This system is designed to be quickly demountable. For this reason, rather than fixing the threaded rods to the lengths of conduit, we provide a hole and slot to enable them to be "clipped" into place (Figure 7-18).

To make this hole and slot, we use regular drills and a vice. First drill the large hole. Pick a drill bit that will give you a hole that is larger than the nut which you glued onto the end of the threaded rod. Drill the hole about an inch higher than the lines that you marked on the conduit. With the conduit clamped securely in a vice or workbench, drill the hole.

Next we move on to the slot. It would be nice to be able to mill this slot, but in the real world, not many hobbyists have access to milling machines, so instead, just pick a fairly decent drill bit. The drill bit should be ever so slightly smaller than the threaded rod you are using, so that as you slide the rod into the slot, the slot grips the rod. Starting with the drill bit in the hole you just made, push on the drill sideways towards the lines that you drew on the conduit.

You will end up with three (or four) identically slotted pieces of conduit as in Figure 7-19. All the holes and slots should line up. You will end up with six (or eight) holes and slots in total, two on each piece of conduit.

For the three-finned launcher

We now need to make the aluminum angle brackets. To do this we will need some meter length of aluminum

Figure 7-19 *The finished set of conduits.*

right-angle. Figure 7-20, shows how the angle needs to be marked up. We will be taking cuts from the aluminum at the intervals shown, which are 60° from one of the sides of the angled stock. Figure 7-21 shows the angle being marked up with a marking gauge. Once the angles have been cut, we can use the remaining side to fold at the point of the angle. We will be creating a hexagon with two short sides and one long side. The short sides will have a hole drilled in the centre. This hole should be bigger than the threaded rod to allow the rod to be slid in and out of the hole easily.

Note that at one end of the aluminum we need to remove a tab of metal. The remaining tab is folded around (see Figure 7-22), and a small hole drilled through both the tab and the piece of aluminum it overlaps. A small self-tapping screw can then be used to secure the assembly (see Figure 7-23).

The final assembly will resemble Figure 7-24.

For the four-finned launcher

Building a launch pad for a four-finned rocket is in some ways much simpler than for the three fins. Rather than the complex angles, all you need is a 45° miter block. Mark the aluminum out to give four equal lengths and a small tab, then remove the triangles from the positions where the angle needs to be bent.

Making the launch pad supports

The structure needs to be as rigid as possible. In addition to the rigidity of the parallel plastic conduit lengths, we supplement the strength of the launcher by three parallel metal bars. These are "sandwiched" together with the aluminum hexagons, and are held in

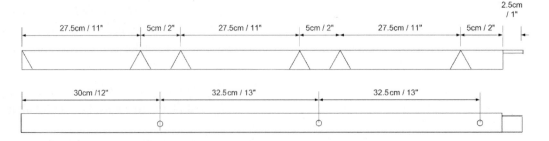

Figure 7-20 *Measurements for the aluminum brackets.*

Figure 7-21 *Marking out the aluminum angle.*

Figure 7.23 *The tab of metal secured with a self-tapping screw.*

Figure 7-22 *The overlapping tab of metal used to secure the hexagon.*

place by the wing nuts that are used to adjust the position of the threaded rods. The supports are simply a piece of flat metal bar with a hole in one end and a 90° bend (to allow that end to be bolted to the base), and two further holes drilled in the same position as the holes in the plastic conduit.

Assembling the launcher

You now need to assemble the hexagons and supporting angle with the threaded rods. Figure 7-25 shows the

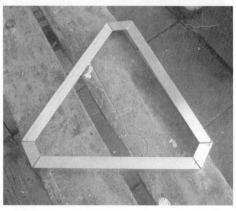

Figure 7-24 *The hexagonal assembly viewed from below.*

assembly test-fitted with some plain rod. You will now need to get together the wing nuts, washers, and plain nuts that suit the threaded rod, and assemble them in accordance with the part and assembly drawing in Figure 7-26. In addition to the mandatory nuts and bolts, I also like to include a small bead. This is just a small plastic bead from a child's necklace set, which slides over the rod. This makes the job of aligning the rods much easier, as the bead can be slid to the position where the nut is desired, and then the nut wound to this position – setting the position first and then winding the nuts up to this position is far easier than trying to

remember what position all the others are set to. The bead can be seen more clearly in Figure 7-27.

Now you can trial assemble your gantry launcher. The easiest way is to start by assembling the aluminum hexagons to the support brackets, shown in Figure 7-28 using the threaded rods as a fixing method. Set all the distances to the same amount – this way it is easier when it comes to clipping on the lengths of conduit. Now clip the conduit onto the ends of the threaded rods (where the nuts have been glued in place), and tighten the nut and washer against the slot in the conduit to secure the conduit in place. It will look something like Figure 7-28.

Ensure that all the three tubes are parallel and aligned. This is shown in Figure 7-29.

The support brackets must be bolted to the base; this is simply a matter of drilling a hole in the plywood, and using a sturdy nut and bolt to secure the support brackets. See Figure 7-30.

The launcher is now ready to use! Your final masterpiece should look like Figure 7-31.

This launcher is a little bulky in its fully assembled form; as a result, it is best to dismantle it for transport. It has been designed to be easy to assemble and disassemble. As can be seen in Figure 7-32, the whole launcher assembly EASILY fits into a small car; a Ford Fiesta is shown.

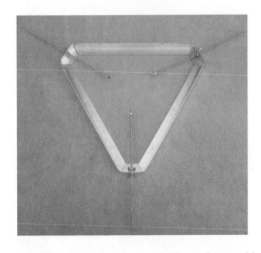

Figure 7-25 *The hexagon assembly trial assembled with plain rods.*

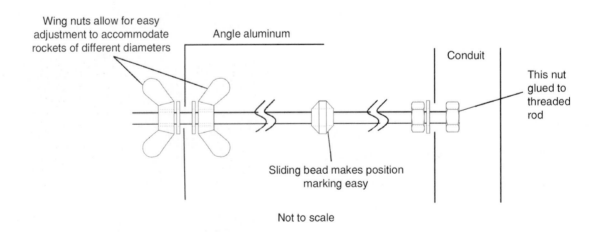

Figure 7-26 *Part and assembly drawing for gantry launcher rods.*

Figure 7-27 *Marker bead on threaded rod.*

Figure 7-28 *Aluminum hexagons bolted to support brackets using threaded rods.*

Figure 7-29 *All three tubes are aligned.*

Figure 7-30 *The support brackets bolted to the base.*

Figure 7-31 *The completed launcher.*

Figure 7-32 *Transporting the launcher to site.*

Project 24: Build a "Rail" Launcher

You will need

- Two lengths of C section plastic/metal (> 2 m)
- 1 ½ × 2 in (3.5 × 5 cm) planed sawn softwood
- Panel pins
- Plywood
- Metal bar stock 1½ in (3.5 cm) wide (aluminum/ mildsteel)

Tools

- Hammer
- Drill
- 1 mm drill bit
- 8 mm drill bit
- Jigsaw
- Screwdrivers
- Hacksaw

In this project we are going to build a rail launcher. A rail launcher differs from a rod launcher in that instead of requiring a piece of tube glued to the side as a launch lug, it requires a small plug or pip, which runs along a rail (hence the name rail launcher). Rail launchers are suitable for larger, heavier rockets, where a rod launcher might be inappropriate. Rail launchers still have the disadvantage of a prominent lug; as such it will be a cause of aerodynamic drag.

Launch rail

Once you have assembled the parts in the list above you can begin. To start, you need to take the two lengths of C section plastic/metal, and drill a series of 1mm holes on a 6 in spacing at the back of the C section, as shown in Figure 7-33. If you are using metal, the holes are essential; if you are using plastic, the holes prevent the plastic from buckling when we fix it to the rod with pins.

Figure 7-33 *Drilling the plastic C section to accept pins.*

Figure 7-34 *The profile of the rail launcher.*

Next take a length of the sawn and planed timber, and using the pins, nail one of the C sections to the short edge of this piece. Using a ruler, measure a gap that is appropriate to the launch lugs that you will be using; ¼ in (5 mm) is quite common. Now extend this line along the length of the timber. It is essential that you mark this up, as it is imperative that a constant gap is provided along the length of the launch rail to prevent fouling. The profile of the launch rail will look something like that in Figure 7-34 when you are finished.

We now need to add some additional wood at the back of the rail near the base (Figure 7-35). This is because we want our rail launcher to accommodate as wide a range of rockets as possible. If we were to mount the rail straight to the base using brackets, we would run the risk of fins snagging on the base. For this reason, a little extension allows us to use a wider range of rockets. This is simply drilled and screwed to the existing piece of wood.

Metal brackets

Now that our rail is prepared, we need to think about making the metal brackets for the base. This is a really simple affair, and can be made using a piece of 1½ in (3.5 cm) flat bar. If you have engineer's blue, paint and mark out four identical bars in accordance with Figure 7-36, if not, you can use permanent marker.

Figure 7-35 *Adding wood to the back of the rail.*

These brackets need to be as identical as possible, but millimeter accuracy is not imperative.

The bracket now needs to be bent in accordance with Figure 7-37. This can be done in a vice by hand with little problem.

You should now have four identical metal brackets (Figure 7-38). It is a good idea to paint them if they are made from steel, as if left unpainted, you will find that they quickly go rusty. Aluminum can be left unpainted however.

Launch angle set

We now move on to making the launch angle set. This allows us to vary the launch angle 30° from the vertical. It comprises two sheets of plywood, with a semicircle marked atop of a 3 in (7.5 cm) rectangle. The rectangle gives us a little space to clear the launch rail. The centre of the semicircle will be your first hole, from here measure 5 in (12.5 cm) using a compass, and draw an arc. All of the holes for our "angle presets" will lie along this line. Now, using a protractor, mark out divisions of 5° from the vertical and extend these lines to meet the arc. Now drill all the points where the lines cross (see Figure 7–39). Only mark angles within 30° of the vertical.

Now cut out the semicircles, and using a bit of the planed softwood, cut two lengths and mount these to the base of your semicircles. They should resemble those in Figure 7-40.

Assembling the launch rail

Now the brackets need to be screwed to the launch rail. You should fix the first one to the very bottom of the rail, and the second bracket so that the center of the hole that fixes to our launch angle set is 5 in (7.5 cm)

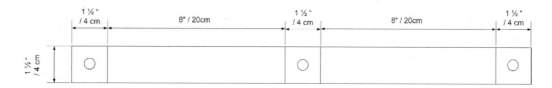

Figure 7-36 *Cutting and drilling diagram for metal bar.*

Figure 7-37   *Bending diagram for metal bar.*

Figure 7-38   *Four finished metal brackets.*

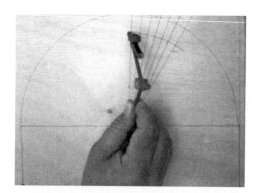

Figure 7-39   *Marking out the launch angle set.*

from the bottom bracket's corresponding hole. Remember we marked out all of our adjustment holes, 5 in (7.5 cm) from the center hole.

Now using four nuts and bolts, bolt the bottom brackets to the center holes of the launch angle set, and the bolt the upper brackets to the most vertical hole of the launch angle set.

Figure 7-40   *Two cut, drilled, and finished launch angle sets.*

Figure 7-41   *The completed base.*

Moving this bolt allows you to adjust the launch angle

Figure 7-42   *Adjusting the angle of launch.*

Figure 7-43 *The finished launcher.*

You can now position your launch rail and angle set over the blast deflector and launch pad. Drill at least two holes in the batons on each launch angle set, and then bolt these to the base. The completed base will look like that in Figure 7-41.

Adjusting the angle of launch

Adjusting the angle of launch is now a very simple affair; the top bolt is removed from the launch angle set, and the launch rod is moved so that the holes in the brackets correspond to one of the other presets (Figure 7-42). Now insert the bolt, ensuring that you remember to replace the nuts to secure the bolts and prevent them from working loose. The nuts do not need to be tightened excessively, their main function is to prevent the bolts from falling out.

Now admire your accomplishment! The finished launcher is shown in Figure 7-43.

Project 25: Launching Your First Model Rocket

You will need

- Launcher (rod/rail/gantry)
- Launch controller
- Safety key
- Model rocket(s)
- Rocket motors
- Igniters
- Recovery wadding
- Batteries

Before you can fly a model rocket you need to find a suitable flying field. Check your intended site against the checklist in order to ensure its suitability. Furthermore, you will need to check its size against Table 7-1) to ensure that it is suitably large.

Warning

Ensure that your site is free from trees, obstructions, or powerlines. Never fly your rocket near to powerlines as shown in Figure 7-44; it is all too easy for the recovery system to become snagged in a pylon!

Choose a clear area, and unpack your launch pad and assemble it. If there is dry grass on the ground, consider spreading out a sheet or small tarpaulin to catch any falling debris and prevent it from igniting the grass.

Unravel your lead, and set up your launch controller well out of harm's way. You should be at least 15 feet (5 m) away from the launcher for anything up to a type E motor (Table 7-2). Make sure that the safety key is in your pocket and that the launch controller is not armed.

Table 7-1

Model Rocket Field Sizing

Motor Size	Minimum Field Dimensions Square Feet	Minimum Field Dimensions Square Metres
1/2A, 1/4A	50	15
A	100	30
B	200	60
C	400	120
D	500	150
E	1000	300
F	1000	300
G	1000	300
H	1500	450
I	2500	760
J	5250	1600
K	5250	1600
L	10500	3200
M	15500	4700

Figure 7-44   *Snagged in a pylon.*

Retract the launch clip and load your model rocket with a motor. Remove any protective device from the end of your launch rod. Now slide your rocket onto the launch platform. Insert an igniter as per the instructions earlier in this chapter. Retreat to the launch controller.

Now give a loud audible countdown. We miss out "five" because it sounds similar to "fire". 10... 9... 8... 7... 6... ... 4... 3... 2... 1... Press the button! The scene should look something like Figure 7-45.

A word on using copperhead igniters

If you are launching a model that uses a reloadable composite motor, then the chances are that you will be using a copperhead igniter. These are a little different from your traditional model rocket igniter as they look like a single piece. Basically there are two copper strips either side of an insulating core, at the end of the igniter is a conductive pyrogen which ignites the propellant.

Because the igniter is a single piece rather than a conventional igniter with two leads, you have to insulate one jaw of each crocodile clip, using masking tape for example, and then clip the leads on to the igniter so that one insulated side is on each side of the igniter and the opposing conductive side is on the other.

When rockets turn nasty... what can go wrong... "CATO"

CATastrOphic failure

> "...not now CATO!" exclaimed the rocketeer (after Peter Sellers in the *Pink Panther* films).

CATOs happen unexpectedly and without warning. They can be terrifying (see Figure 7-46), and and those who are in the vicinity can be seriously injured if not standing the correct distance away.

Table 7-2
Safe distances from the launcher

Motor size	Minimum	Distance
1/2A, 1/4A,	7 feet	2 metres
A	7 feet	2 metres
C	10 feet	3 metres
D	10 feet	3 metres
E	16 feet	5 metres
F	23 feet	7 metres
G	33 feet	10 metres
H	33 feet	10 metres
I	49 feet	15 metres
J	148 feet	45 metres
K	148 feet	45 metres
L	197 feet	60 metres
M	295 feet	90 metres

Figure 7-45 *Watch with awe and wonderment as your creation takes off! Images courtesy Peter Barratt.*

CATOs happen when the propellant mixture has been thermally cycled, that is to say repeatedly heated and cooled. As a result cracks can form in the motor propellant grain, causing it to burn at different rates, and ultimately explode.

If you are standing the distance away from your rocket recommended in the safety code (Table 7-2), then you will be fine.

In terms of preventing CATOs, model rocket motors are a sealed unit and not suitable for alteration; the best thing to do is to use fresh motors, and store them at a constant temperature.

Figure 7-46 *Catastrophic launches.*

Project 26: More Advanced Launch Techniques

You will need

- Clustered or multistage model rocket motor
- Multiple motors
- Multiple igniters
- Copper wire

Tools

- Side cutters

This project deals with launching rockets which have multiple motors and/or stages. For these models, a little bit more thought has to go into how you are going to ignite all of the motors simultaneously.

First of all, we always connect in parallel; this is because if we were to connect in series, the circuit would be broken as soon as one igniter burnt through. Always use the "advanced launch controller" and a large battery, as multiple igniters draw a large current.

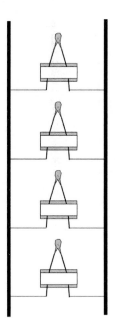

Figure 7-47 *Igniters in parallel.*

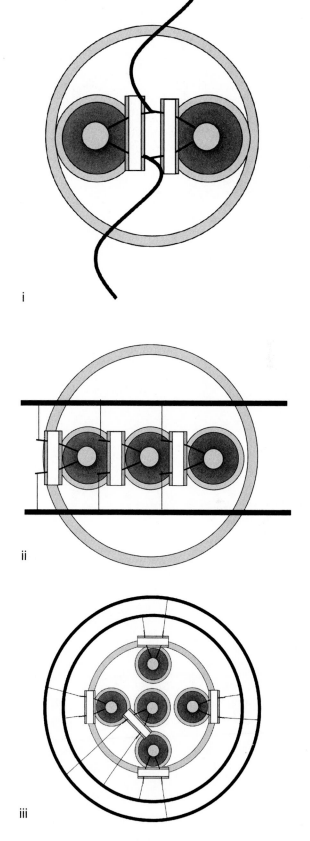

i

ii

iii

Figure 7-48 *Wiring igniters for cluster motors.*

Hint

Always use your BEST igniters with the most amount of pyrotechnic material in them to ensure successful multiple motor launches.

Note

When igniting multiple motors, ALWAYS connect the igniters in parallel, NOT in series. See Figure 7-47.

For simple rockets with only two motors, it is often just a matter of using the igniter wires with a pair of crocodile clips to connect to the launch controller (see Figure 7-48i).

For slightly more complex motor arrangements, you will find that the igniter leads are not long enough to stretch to the crocodile clips. In these instances, use two pieces of copper wire as "bus bars"; these will conduct the electricity to all igniters simultaneously. They can either be a pair of straight wires, as in Figure 4-48ii or two concentric circles around the perimeter of the rocket motors to which all of the igniters attach, as in Figure 4-48iii.

Chapter 8

Rocket Math

So you wanna call yourself a rocket scientist (part 2)...

If you are going to do things properly, you really need to get to grips with "rocket math." You will be a far more competent model rocket scientist if you understand the math that underpins the science behind making your models fly.

Think of it this way. If you invest the time and effort in learning how to calculate how your model flies, then the chances are you will be able to build a model that flies right first time, recovers like a dream, and can be reused and reflown many times. If you don't invest the time in the math, then you can only learn by trial and error; this approach could cost you dearly in the long run.

We're going to start things nice and easy.

All you will need for these exercises is a pen, paper, and decent scientific calculator. We are going to be dealing with some fairly serious maths, so your standard cheap "plus, minus, times, divide" calculator isn't going to cut the mustard! You want one with a decent array of functions (for which read the more buttons the better!).

As you get more used to these equations, you might want to procure a calculator that allows you to save the formulas as presets. Then it is simply a matter of punching in the numbers in response to a series of prompts for each variable. In the long run, if you are planning to build a lot of rockets, a calculator with this function is incredibly useful.

A graphic calculator is VERY useful, but not essential. In rocketry, most activities take place over a period of time, and it is nice to be able to plot graphs in order to represent things visually and make them easier to understand.

Of course posh tools are all well and good, but a pencil and graph paper works equally as well for plotting graphs – learning how to plot a graph is a good skill, and there is no harm in practising!

Computers make things incredibly easy now; they are such a powerful tool, and should not be overlooked by the model rocketeer. For commonly used equations, it is nice to be able to create a proforma spreadsheet with the formulas already punched into the boxes, so that when you need to work something out, you simply open up your template and bash in the number! This will certainly relieve some of the tedium of repetitive calculation. Microsoft Excel is widely accepted and used by many people; however, if you are on a budget, Star Office and Open Office are both well worth looking into and can quickly be found by an internet search.

Project 27: How Fast? How Far? – Getting to Grips with Basic Math for Physics

You will need

- Pen
- Paper

Tools

- Calculator
- Spreadsheet (optional)

You read all about "Newton's Laws" in Chapter 2: Rocket Science. Well, we are now going to look at the underpinning math. These laws are pretty good until you approach the speed of light. Luckily, even the best rocket motors on the market these days aren't going to achieve warp speed, so the underpinning maths here should hold for all our "impulse power" rockets.

---

s = distance

Unit: m (metres)

A Standard International System of Units.(SI) unit of length.

v = velocity

Unit: m/s or m s^{-1} (metres per second)

Both formats mean the same thing; the minus sign is a bit like moving the dividing line and putting it "up in the air," if that helps you understand it any better. By saying metres per second, we are saying that our rocket travels "a number of meters" in any given second.

v_1 = velocity at start of given period

Unit: m/s or m s^{-1} (meters per second)

v_2 = velocity at end of given period

Unit: m/s or m s^{-1} (meters per second)

V_{av} = average velocity over given period

Unit: m/s or m s^{-1} (meters per second)

a = acceleration

Unit: m/s^2 m s^{-2} (meters per second squared)

This is like saying metres per second, per second.

t = length of period

Unit: s (seconds)

$$S = vt \qquad (1)$$

This equation says that the distance traveled is equal to the velocity (speed) multiplied by the amount of time. If your car is traveling at a constant velocity of 30 miles per hour (mph) and you travel for 2 hours, then you will have traveled 60 miles. Relating this to rocket science, if your rocket travels at a constant velocity of 200 m/s, and it travels for 4 s, then it will have travelled 800 meters. Capisce?

$$V_{av} = \frac{(v_2 + v_1)}{2} \qquad (2)$$

This equation is saying that the average speed of our model rocket is equal to the average of the start and finish velocities (the velocity at the start of a given period plus the velocity at the end of that period, divided by two). This of course assumes that the acceleration is at a constant rate.

For example, if we are driving along the road at 30 mph, and we accelerate steadily to 50 mph, the average speed over that period of constant acceleration would be equal to 40 mph (30+50/2).

A rocket example: if your rocket starts at a speed of 0 m s^{-1} on the launch pad, and the rocket accelerates to 300 m s^{-1} at a constant rate, then the average speed over that period will be 150 m s^{-1}.

Next let's look at what happens if we want to find the acceleration of our rocket over a given time. Look at equation 3.

$$a = \frac{(v_2 - v_1)}{t} \qquad (3)$$

Equation 3 tells us that if a rocket is accelerating at a constant velocity, that acceleration will be equal to the final velocity minus the start velocity, divided by the time.

By now you should be getting the hang of these motion equations; here's a couple more for your delectation.

If we want to find the distance traveled, and we know the start speed, the acceleration, and the time taken, we can use the following equation:

$$s = \frac{v_1 t + (at^2)}{2} \qquad (4)$$

We could use it, for instance, to find the distance that an object falls when we know its initial speed (assuming it is from rest "0"), the acceleration due to gravity, say 9.8 m/s, and the time taken. What good is that to us as model rocketeers? Well, you might have seen the Estes "Max Trax." Essentially, the microcontroler inside is doing something similar. It knows the speed at which it will fall with any given parachute. It starts from a velocity of zero, and by timing the period from ejection to "touchdown" it is pretty easy to work out roughly how high the rocket went.

$$2as = v_2^2 - v_1^2 \qquad (5)$$

Finally, and this should seem kids' stuff by now, the acceleration multiplied by the distance, then multiplied by two, is equal to the final velocity minus the start velocity.

If you have mastered the above, you can now go and annoy your family with some REAL rocket math!

Online calculators

If by now you are hopelessly lost and need a little guiding hand in the dark, here are a number of online calculators that make the business of calculating distance, speed, time, and acceleration much easier. Try calculating a sum and working through it in order to improve your knowledge. You can do it yourself, all it takes is a little persistence.

www.gazza.co.nz/distance.html

www.csgnetwork.com/csgtsd.html

www.projects.ex.ac.uk/trol/scol/ccacceln.htm

Project 28: Calculating How High Your Rocket Went

We are going to take a little break from the plain algebra – I know you are going to be REALLY disappointed. Instead, we are going to apply a little geometry!

Rocketry height estimation is all about geometry and triangles. Essentially, when we try to determine how high our rocket went, we are using the laws of geometry to solve triangle problems.

Calculating how high your model went can be done by taking some angular measurements from the ground; by combining them with linear measurements that you already know, you can work out the other angles and lengths of the triangle.

We are going to start off with some easy examples, and work our way we towards some harder problems. you might wonder why we use a more complex method when a simple one will do. The answer is accuracy. The more effort we put into the tracking and maths, the more accurately we can determine the position of our rocket.

All the methods that are presented here are "elevation only"; that means that you can use a really simple tracker that simply measures the angle between the ground and your rocket in flight. There are some more complex methods available, which require not only the angle of elevation, but also the "rotational angle," known as the azimuth. If you are having trouble visualizing what I mean, think of the azimuth as how far you have to turn to see your rocket in clear line of sight. You might also like to think of it as a "bearing" in navigational terms; how far you would have to rotate to change direction.

Methods that require both the elevation and the azimuth require slightly more sophisticated tracking devices.

Single-station tracking

In some model rocket fraternities, you will hear this method called "elevation angle only tracking."

Single-station tracking really is pretty simple. We stand a fixed distance away from our rocket launch pad. Follow the rocket visually and, using a device called a theodolite, measure the angle between the ground and the apogee of the rocket. There are a number of ways you can do this. Some simple commercially available trackers are available, the Quest "Skyscope" is one and the Estes "Altitrak" is another.

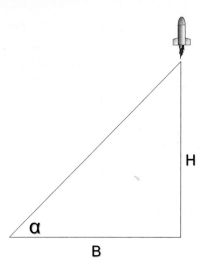

Figure 8-1 *Diagram showing geometry of single-station tracking.*

> Let's define some more terms:
>
> H = calculated height attained by rocket
>
> α = angle of rocket's ascent
>
> B = length of baseline

$$H = \tan \alpha \times B \qquad (6)$$

For a diagrammatic view of the setup look at Figure 8-1.

This method is a little flawed because it assumes that our rocket makes a perfectly vertical flight. This is rarely the case! To improve the accuracy of your results, the best thing that you can do when tracking is to position yourself at right angles to the wind movement in relation to the launch site.

Many older rocket books refer to "tables of tangents"; while these do not require batteries, is is just as easy to carry with you a scientific calculator.

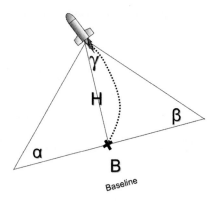

Figure 8-2 *Diagram showing geometry of dual-station tracking.*

$$H = \frac{B \times \sin \alpha \times \sin \beta}{\sin \gamma} \qquad (7)$$

This tracking method if more accurate if you position the trackers at opposite sides of the launch pad, in line with the direction of the wind.

Dual-station tracking

Here is a slightly more accurate method that employs two stations; again, you just require a simple angular measurement. We know that the interior angles of a triangle will add up to $180°$, therefore it is quite easy to calculate the angle between the two trackers. The baseline is measured using a surveyor's tape. From these values the height that the rocket attained is easy to compute.

> H = calculated height attained by rocket
>
> B = length of baseline
>
> α = (1st tracker) angle of rocket's ascent
>
> β = (2nd tracker) angle of rocket's ascent
>
> γ = angle between both tracker's angles

Increasing accuracy

If you want to explore increased tracking accuracy, take a look at the following program, which provides an easy way of estimating the altitude attained by your rocket. It integrates data from up to 12 stations, simply by entering each station's bearing and distance in relation to the launch pad.

http://nmwg.cap.gov/santafe/Activities/StartAltiCalc.htm

Project 29: A Simple Tracker

You will need

- Photocopy of Figure 8-3
- Card
- Glue
- Paper fastener
- Lump of plasticine
- Scalpel

Tools

- Scissors
- Hole punch

In this project we will be making a simple elevation-only tracking device.

You need a photocopy of Figure 8-3. Stick this to some card, and carefully cut out around the black lines. Now fold the tracker body along the centre line and glue it to together itself.

You will now need a hole punch. Punch the single hole in the tracker body, and the two holes in the tracking arm. Now with a scalpel, carefully cut out the window in the tracking arm. This is the shaded area.

Assembly of the tracker is simple. Fold the tracking arm around the tracker body ensuring that the numbers can clearly be seen through the window. Fasten the assembly together by inserting the paper fastener through the punched holes, and weight the bottom of the tracker arm with a small lump of plasticine.

Hold the tracker in your hand, and raise it to the sky. As the tracker arm is weighted, its natural tendency is to drop to the ground. Read off the number in the centre of the window, or if the angle is between two of the

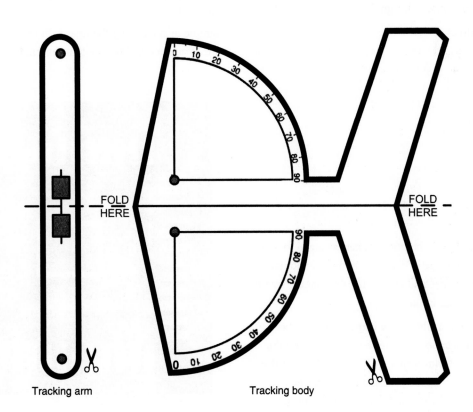

Figure 8-3 *Tracking device template.*

marked values estimate, its value using its position relative to the two adjacent angles as a guide.

Use this technique when tracking rockets.

Ready-made trackers

For something a little more professional to track your rockets with:

The Estes Altitrak, available from:
www.estesrockets.com

Apogee Altitude Tracking Device:
www.apogeerockets.com/altitude_tracker.asp

Quest Skyscope
www.questaerospace.com/pages/products_new.htm

Project 30: Calculating How High Your Rocket Went from an Aerial Photograph

You will need

- Model rocket camera – this could be

- The digital camera design featured in this book
- Estes Astrocam
- Estes Snapshot
- Estes Camroc

- Launch vehicle

Tools

- Vernier calipers

We can also gauge the height a rocket managed to attain by looking at an image of features on the ground, taken at the rocket's apogee and comparing their actual size to the size they appear in the image.

If you are using a film-based camera such as the Astrocam, you need to measure the image size on the negative. The 110 film format is 13 mm × 17 mm; if you get some prints made, take a piece of plastic and draw a grid with 17 vertical divisions and 13 horizontal divisions to fit the print. This will enable you to work

out the size of each feature on the negative, with each square corresponding to a millimetre square on the negative or thereabouts.

If on the other hand you are working with a model rocket digital camera, such as that described in the pages of this book, then unfortunately, things aren't quite as simple as measuring the image on the screen of your PC. You need to imagine that that image was recorded

Figure 8-4   *Aiptek PenCam CCD.*

on a tiny charge-coupled device (CCD) sensor which is probably much smaller than your little fingernail. (see Figure 8-4).

$$H = \frac{O_{\mathrm{L}}F}{I_{\mathrm{L}}} \qquad (8)$$

H = height of rocket camera from ground

O_{L} = object length on ground

I_{L} = image length on image sensor/negative

F = focal length of lens used

Project 31: Calculating the Size of an Object on the Ground

You will need

- Model rocket camera – this could be

  - The digital camera design featured in this book

  - Estes Astrocam

  - Estes Snapshot

  - Estes Camroc

- Altimeter or tracking device
- Launch vehicle

A situation might arise whereby you want to know the size of a large building, the distance between two points, or the size of an area.

There are number of ways of calculating this. The most simple method is if you have a picture that has been taken vertically, not obliquely, and you know the size of another object on the ground. You can then use ratios to work out how big the object is by looking at the difference between their respective sizes on the print.

Let's work through an example:

We have a picture that includes a car that we know to be 4.5 m long, and a warehouse which is of an unknown size.

We carefully measure the picture with our vernier calipers, and find that the warehouse measures 6 cm long on our print, and our car measures 3 mm. The 6 cm (60 mm) is 20 times the 3 mm; therefore, our warehouse is 20 times bigger than the car. As we know that the car is 4.5 m long, we multiply the 4.5 by 20 to give us 90 m. Simple!

What if we don't know the size of anything in our picture? We need some other fixed point of reference from which to perform our calculations. This isn't a problem, as long as we know the height of the rocket when it took the picture. This could be done using either the altimeter in the model rocket flight computer's device, or a commercially bought altimeter. Failing that, we can track the rocket using the tracking device we made earlier (Project 29: A Simple Tracker) and approximate the height at which the picture was taken at.

This time we use a rearrangement of the equation used in the previous project (see equation 9)

$$O_{\mathrm{L}} = \frac{I_{\mathrm{L}}H}{F} \qquad (9)$$

Project 32: Calculating the Measurements for a Boat-tail Transition

You will need

- Stiff card, plastic or plasticard

Tools

- Compass
- Pen
- Calculator

Often we need to lay out a transition between two tubes of dissimilar diameters. It is possible to do this with a "solid" transition, such as one turned from balsa wood, but often for reasons of weight-saving or access, a hollow transition would be better. This could be because we want to put a "boat tail" on the bottom of our model to improve the aerodynamics, or because we need to accommodate a motor in the middle of the transition.

Unfortunately, it isn't simply a matter of just cutting out a strip of paper and bending it a bit – we need to calculate the geometry of the transition in order to make sure that what we cut out will fit.

Look at Figure 8-5. Start with the left-hand illustration, looking at your transition sideways on. The dimensions that you really should be thinking about are d_1, d_2 and L. They are your smallest tube diameter, largest tube diameter, and the length over which you wish this transition to occur. Once you have ascertained these dimensions, it is possible to calculate r_1, r_2, and θ. Work through equations 10 to 14. These are the dimensions you will need to transfer onto something that resembles the image on the right. Once you have done this, you can cut out your transition and fold it like a cone.

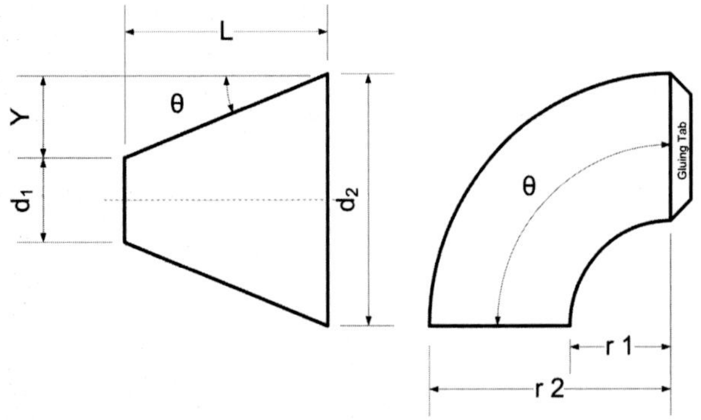

$$\gamma = \frac{d_2 - d_1}{2} \qquad (10)$$

$$\sin \theta = \frac{1}{\sqrt{\left(\frac{L}{\gamma}\right)^2 + 1}} \qquad (11)$$

$$\phi = 360 \sin \theta \qquad (12)$$

$$r_1 = \frac{d_1}{2 \sin \theta} \qquad (13)$$

$$r_2 = \frac{d_2}{2 \sin \theta} \qquad (14)$$

You can either cut out an additional little tab of material and stick this onto the back of the opposing piece of material, or you can cut out a strip of material and stick it on the back; this is a less obtrusive method, although a little trickier. Hold everything temporarily in place with a little masking tape until the glue sets.

And that's it. You are ready to go!

Figure 8-5 *Critical transition dimensions.*

This is where things start to get a little bit heavy with some pretty hairy mathematics. Just bear in mind that it is all just algebra in one form or another – break everything down into smaller bite-sized chunks and you will find it all a lot easier.

The Barrowman equations are named after James and Judith Barrowman, who perfected the method in March 1967. James S. Barrowman worked for NASA's Sounding Rocket Branch, and submitted his Master's thesis *The Practical Calculation of the Aerodynamic Characteristics of Slender Finned Vehicles* to the Catholic University of America.

This method calculates the centre of pressure, and results in a measurement from the tip of the nose. Relate your model to the diagram shown in Figure 8-6.

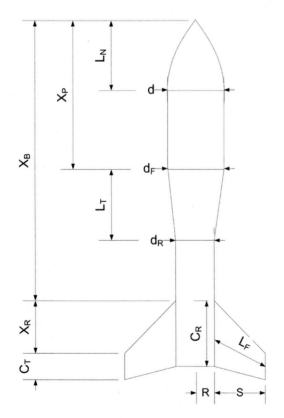

Figure 8-6 *Barrowman equations diagram.*

- For a cone-shaped nose, use equation 15.

$$\overline{X}_n = \frac{L}{3} \qquad (15)$$

- For an ogive-shaped nose where the length is less than 6 times its radius, use equation 16.

$$\overline{X}_n = 0.534\,L \qquad (16)$$

- For a parabolic nosecone, use equation 17.

$$\overline{X}_n = \frac{L}{2} \qquad (17)$$

Defining terms:

d_F = diameter of front of transition (d_2 in the last project)

d_R = diameter of rear of transition (d_1 in the last project)

L_N = length of transition (L in the last project)

X_P = tip of nose to front of transition

d = diameter at base of nose

C_R = fin root chord length

C_T = fin tip chord length

S = fin semispan

L_F = length of fin midchord line

R = radius of body rear end

X_R = fin root leading edge to fin tip leading edge

X_B = nose tip to fin root chord leading edge

$$(C_N)_T = 2\left[\left(\frac{d_R}{d}\right)^2 - \left(\frac{d_F}{d}\right)^2\right] \qquad (18)$$

$$X_T = X_P + \frac{L_T}{3}\left[1 + \frac{1 - \dfrac{d_F}{d_R}}{1 - \left(\dfrac{d_F}{d_R}\right)^2}\right] \qquad (19)$$

- For a three-finned model, use equation 20.

$$(C_N)_F = \left[1 + \frac{R}{S+R}\right] \left[\frac{12\left(\frac{S}{d}\right)^2}{1 + \sqrt{1 + \left(\frac{2L_F}{C_R + C_T}\right)^2}}\right]$$

$$\left[1 + \frac{R}{S+R}\right] \left[\frac{16\left(\frac{S}{d}\right)^2}{1 + \sqrt{1 + \left(\frac{2L_F}{C_R + C_T}\right)^2}}\right] \quad (20)$$

- For a four-finned model, use equation 21.

$$\overline{X}_F = X_B + \frac{X_R}{3}\frac{(C_R + 2C_T)}{(C_R + C_T)}$$

$$+ \frac{1}{6}\left[(C_R + C_T) - \frac{(C_R C_T)}{(C_R + C_T)}\right] \quad (21)$$

$$(C_N)_R = (C_N)_N + (C_N)_T + (C_N)_F + \dots \quad (22)$$

$$\overline{X} = \frac{(C_N)_N \overline{X}_N + (C_N)_T \overline{X}_T + (C_N)_F \overline{X}_F}{(C_N)_R} \quad (23)$$

Project 34: Recovery Device Calculation

In this little exercise, we are going to size some parachutes correctly in order that your model makes a safe descent and does not fly away cover the treetops.

To do this, we are going to have to explore drag a little bit (see equation 24).

$$D = 0.5 \, \rho V^2 C_d A \quad (24)$$

Defining terms:

D = drag force

ρ = density

C_D = drag coefficient

A = area of the parachute

V = velocity

Density is the density of the gas that our parachute is dragging through; in this case, the gas is air, and we can take the density to be 1.22 kg/m^{-3}

This is all fine if all we want to calculate the drag force exerted by a given parachute, but what if we want to calculate the diameter of the parachute needed to slow our model to a sensible speed?

We need to do a little more groundwork. First of all, we need to determine the speed at which we want our parachute to hit the floor, and how much our model weighs. We can combine the above with a bit of basic physics: force = acceleration × mass. My physics teacher Mr Spary used to drum into us that the way to remember this was Forces Are Magic, the F, A, and M sitting neatly in a magic triangle. We can then combine this with a bit of geometry, assuming that we want a circular parachute.

This leads us to equation 25.

$$D = \sqrt{\frac{8\,mg}{\pi\rho C_D v^2}} \quad (25)$$

Table 8–1
Common Parachute Shapes and Their Equivalance to a Circular Parachute

Parachute Shape	Measured	Surface area	Diameter of Equivalent Circular Parachute
Circle	Diameter	0.79 × diameter	Equivalent diameter = 1
Square	Across flats	Length squared	Equivalent diameter = 1.1 × length
Square	Across corners	Length squared/2	Equivalent diameter = 0.8 × length
Hexagon	Across flats	Length × 0.87	Equivalent diameter = 1.1 × length
Hexagon	Across corners	Length × 0.65	Equivalent diameter = 0.9 × length

In this equation, D now becomes the diameter of the parachute – this is the one to note.

We now use the following terms:

D = chute diameter in SI units (m)

m = rocket mass in SI units (k)

g = acceleration due to gravity = 9.8 ms^{-2}

π = Pi (3.14 ....)

ρ = density (of air 1.22 kg m^{-3};)

C_d = drag coefficient of the chute

v = velocity (at which your model will hit the ground, ms^{-1})

Drag coefficients

0.75 for a parasheet (flat)

1.5 for a parachute (domed)

Of course, we don't always want to use a circular parachute, sometimes we want to use another shape. Table 8-1 gives some common parachute shapes and their equivalence to a circular parachute.

Online resources

There are some great online tools available:

www.washingtonhighpower.com/
Chute%20Calculator.htm

www.onlinetesting.net/cgi-bin/descent3.3.cgi

Chapter 9

Model Rocket Camera Payload

Aerial and space photography history

Back in 1912, on October 16, the world's first ever aerial reconnaissance mission was executed by a Bulgarian Albatros aircraft. Ever since then, unfriendly nations have been trying to look at each others' activities from the air.

In 1951, a report was published, *Utility of a Satellite Vehicle for Reconnaissance*, which suggested that a television camera in space would be a useful tactical facility for the military. As well as the numerous military applications, it was also quickly realized that imaging from space could be a useful tool in our search for new discoveries.

Tiros (Television and Infrared Observation Satellite), launched in April 1960 was the first camera payload; the satellite returned 22 952 pictures in 78 days.

The first fully successful space reconnaissance mission was performed by a satellite called Discoverer 14 in the week preceding August 25 1960. The mission revealed a picture of 1.5 million square miles of the former USSR.

More recently, the valuable data gained by aerial reconnaissance has been applied by scientists to examine the phenomenon of climate change, crop distribution, urban density and many other factors, all of which are easier to assess from the air than from down on the ground.

Model rocket photography

Camroc

The Estes Camroc, launched in 1965, was the dawn of an age of model rocket photography. For the first time, a kit was available that allowed model rocketeers to take pictures from the sky. The camera accepted circular negatives that had to be processed at home, an awful lot of fuss for your average rocketeer; nevertheless, it broke new ground and started the ball rolling on a path of innovation. Since then Estes have released the Astrocam 110, a camera capable of taking a single shot on standard 110 film. Later branded as the Snapshot, it provides an affordable introduction to the world of "space" photography; however, it still necessitates processing of film, which is costly and a pain.

Project 35: Build a Model Rocket Camera Payload

In this project, we will be building a small payload for our model rocket that is able to take pictures. The difference is, these ones are digital. No more waiting around for prints from the chemist! Simply connect this payload to your PC, download the prints and you are ready to go!

You will need

Aiptek Pencam

Payload rocket (Chapter 5: Constructing model rockets)

Mercury tilt switch

AA battery holder

2 × AA batteries

Tools

Hot melt glue gun

Jeweller's screwdriver

To build this project, you are going to need to get hold of a cheap digital camera. I highly recommend the Aiptek Pencam pictured in Figure 9-1 which is available on eBay new and secondhand for a very reasonable sum of money. The camera is limited in that all memory is internal, and the resolution is only VGA, which may be considered too poor for some applications; however, the design offers a cheap foray into the world of model rocket aerial photography, and as such serves us well in our project.

Caution

This project is destructive! The digital camera used will be torn to bits and reincarnated in a new form! The chances are you will not be able to (or want to) reassemble this camera into its former housing. Take heart from the fact that the camera used for this project is very affordable. This will DEFINITELY invalidate your warranty!

Familiarize yourself with the camera, it can be seen out of the box in Figures 9-2 i and ii. On the front of the camera we see the lens used for taking photographs, and at the top of the camera we see the shutter release button. Turning the camera over, we see the LCD display to the rear, and the "mode" button, which gives us control over the camera's settings. The battery flap is also located on this side at the bottom of the model. The camera will be modified for this project. The supplied

Figure 9-1 *The Aiptek Pencam.*

Figure 9-2i *Aiptek Pencam front view.*

base can be discarded; however, retain the lead and the CD for later use. The manual will also be useful to you as the camera's operation will remain unmodified and the fuctions will still be the same – it is only the power supply and method of triggering the shutter that will be modified.

Figure 9-2ii   *Aiptek Pencam rear view.*

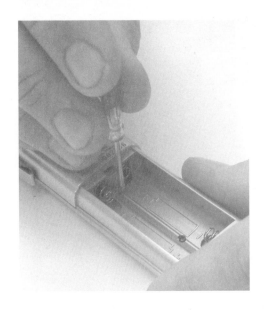

Figure 9-4   *Removing the screws from inside the battery box.*

In order to use the camera, we need to strip it down in order to use its innards; the following is a guide to safe disassembly (without breaking anything!)

Disassembling the Aiptek Pencam

The first step to disassembling the Aiptek Pencam is to remove the battery cover, this does not require any tools and is simply a matter of pushing the cover in the middle, and sliding it off. The battery cover is located to the back of the camera at the bottom. See Figure 9-3 if you are unsure.

There are two small screws within the battery box; they are located in the middle, between where the two batteries go. With the batteries removed, these screws

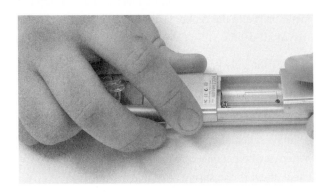

Figure 9-3   *Removing the battery cover.*

can be extracted using a small jeweller's screwdriver. Removal of the screws is shown in Figure 9-4, refer to these pictures to identify the location of the screws.

Once this is done, you should be able to gently lever the case apart with a small screwdriver. The case will feel "stuck" together, but should come apart with only a little persuasion. There are a couple of plastic parts that will come out at this point – the shutter release button and the viewfinder. We will not require these parts at any further stage in the project, so they can be discarded. The area around the USB port is a good

Figure 9-5   *Gently levering the case apart.*

Figure 9-6   *Internal view of the case.*

position to insert a screwdriver and get purchase. This can be seen in Figure 9-5.

The case will now split into two parts. Figure 9-6 shows an internal view.

There are four screws that need to be removed in order to remove the printed circuit board. The first two are located at the base of the printed circuit board near where the spring clips enter the battery compartment. The position of these screws is shown in Figure 9-7. The other two screws are located either side of the lens and go right through the lens, printed circuit board, and through to the case on the other side.

Now is a good opportunity to take a look at the CCD element (see Figure 9-8); be careful not to touch it.

We now need to remove the printed circuit board (PCB) from the case; with the four screws removed, it is simply a matter of using a little bit of leverage to gently prise the PCB from the supporting case. The spring clips that penetrate through to the battery compartment will tend to want to hold the printed circuit board in place. Gently prise them from the battery compartment (see Figure 9-9).

The plastic parts of the case can be discarded, as we will not be needing these any further. We are interested in the PCB, which contains the camera and its support circuitry. You will be left with a couple of loose items. The liquid crystal display (LCD) panel at the back of

the camera will not stay attached, neither will the lens and its two screws, which secure the PCB to the rest of the case.

You will need to replace the LCD at some point. The PCB has a silk screen image of where the LCD should be positioned. Notice that the silk screen is "keyed", that is to say that there is a small notch at one end of the pattern, not unlike the notch used to orient integrated circuits. Notice also, that there is a notch in the glass of the LCD panel that corresponds to the silk screen. (see Figure 9-10).

The LCD is held against the rear of the circuit board by the case when it is shut. Because we are removing the case, we need to find an alternative means of holding the LCD to the printed circuit board. You can epoxy or hot glue the LCD in place, making sure that the epoxy only bonds to non-conducting parts. Keep the metal pads and the rubber strips on the back of the LCD free of all debris.

Once the glue has dried, you can begin to think about the wiring and electronics. We will be connecting a new battery box to the payload and a tilt switch to trigger the shutter. Figure 9-11 shows a simple schematic. Note the orientation of the batteries when connecting them. You can either attach them to the spring clips that are already there – ensuring that you insulate the connection with some heatshrink tubing, or, failing that, you can

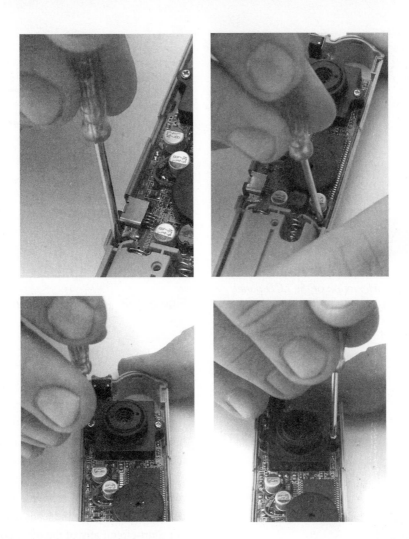

Figure 9-7 *Removing the four screws inside the case.*

Figure 9-8 *The Pencam with the lens removed and the CCD element exposed.*

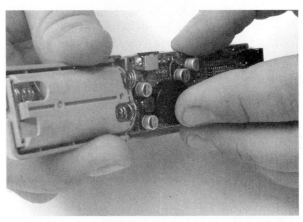

Figure 9-9 *Gently prising the printed circuit board from the case.*

Figure 9-10 *Note the orientation of the LCD panel, and the pip on the left side of the LDC screen.*

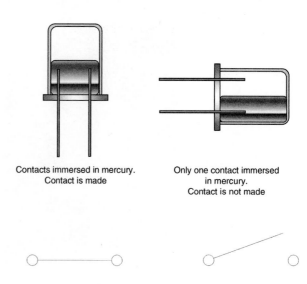

Contacts immersed in mercury. Contact is made

Only one contact immersed in mercury. Contact is not made

Figure 9-12 *Tilt switch operation.*

Tilt switch

Figure 9-11 *Model rocket camera schematic.*

desolder the existing spring terminals, and solder the connections directly to the PCB.

When it comes to attaching the tilt switch, there is little reason to desolder the existing shutter release button, and besides, it provides a handy way of testing the operation of the camera. Instead, "piggy-back" the connections on to the existing shutter button. They are readily accessible and exposed. Again, use a little heatshrink to insulate the leads and ensure that they do not short-circuit any of the other components on the rocket camera PCB. The mode button of the camera is unaffected. Relocating this button would be of negligible value; instead, drill a small hole in the enclosure of the payload, and use a pronged object to poke the button and set the mode. While crude, it is simple and effective.

A note about tilt switches. A tilt switch detects angular movement – once a certain angle has been crossed, a circuit is either made or broken. The tilt switch is a very simple beast to understand. Take a look at Figure 9-12.

The tilt switch is a can containing two terminals. The inside of the can is insulated, apart from the two protruding terminals which are conductive. Inside this tilt switch is a small quantity of mercury. For those of you who were sitting at the back snoring in chemistry lessons, mercury is a metal which is a liquid at room temperature. Being a metal, it also conducts electricity.

The mercury sloshes about inside the can. How it settles depends on the orientation of the switch.

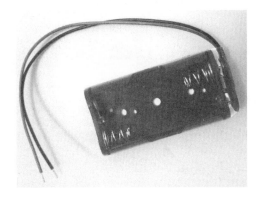

Figure 9-14   *Battery box detail.*

> ## Caution
>
> Never try to disassemble your tilt switch – if it contains mercury, you could be in for a nasty toxic shock!
>
> **Hint:** There is a quick way to tell whether your tilt switch is the mercury or non-mercury type – hold it to your ear and give it a little shake! If you can't hear anything, then it is most likely a mercury tilt switch. If you hear a little rattle, then it is likely to be a non-mercury type. Non-mercury tilt switches have a little ball trapped inside to make the contact instead of the mercury.

In our project, the tilt switch is used to determine whether the rocket is in an upright launch position, a stable flight position, or a horizontal recovery position. Ensure that you buy the right product – the tilt switch required is a low-profile affair such as that shown in Figure 9-13. There are much more expensive tilt switches available that are suited to industrial applications and are encased in large blocks of plastic that can easily be screwed to machinery etc. These are inappropriate for this application and would only add extra weight to our rocket.

You will need a small AA or AAA battery box. When the camera was mounted inside the case, the battery box was integrated into the plastic moulding. While the ingenious, with a little talent, could salvage this battery box, for a few cents it is much easier to buy another! If you can get the type with a 9 V PP3-type battery terminal on the top, then you can readily disconnect your battery, which saves you having to add an

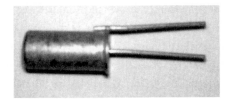

Figure 9-13   *Tilt switch detail.*

ON/OFF switch. The type of battery box that we require for this project is shown in Figure 9-14.

Now you can begin to assemble this disparate band of components into the housing. We will be using the "payload" rocket for our launch vehicle, the method of mounting components is up to you – a mixture of tape, Polymorph, and hot melt glue doesn't make for the most attractive solution, but it is certainly one of the easiest and quickest solutions. Remember when mounting your camera module that you need to be able to remove it intact in order to download the pictures to your PC. Removal of the camera from the housing should not necessitate disconnection of the battery; it should be possible to remove the whole unit, and then plug it in at base camp to a standard USB port using the lead supplied with the PenCam. The USB port remains unaltered.

Figure 9-15 shows the camera mounted inside the payload compartment of our payload rocket. If, after the first flight, you find that the plastic of the payload compartment causes undue reflections or obscuring of the image, you may feel that you want to make a small plastic hole in front of the lens.

To ensure successful deployment of the camera recovery system, it will be separated from the recovery system of the main rocket. The recovery system of the payload rocket deviates from that shown in Chapter 5: Constructing Model Rockets. As can be seen in Figure 9-16, the tube of the payload rocket is first loaded with a plug of recovery wadding, followed by a streamer for the main body. The streamer will tend to cause a faster descent of the rocket body than of the camera; as a result it should fall "in shot," adding

Figure 9-15 *Model rocket camera installed prior to painting.*

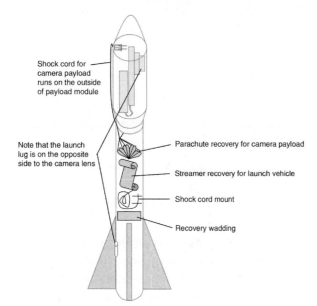

Shock cord for camera payload runs on the outside of payload module

Note that the launch lug is on the opposite side to the camera lens

Parachute recovery for camera payload

Streamer recovery for launch vehicle

Shock cord mount

Recovery wadding

Figure 9-16 *Model rocket camera recovery system.*

interest and variety to your pictures. The streamer is attached to the payload body via a conventional shock cord mount.

The camera payload has its own independent recovery system. A parachute recovers the camera, causing a slow descent. Select the parachute depending on the size of your field, and how fast you want to run! If assembled well, the camera is fairly robust; certainly a hard landing is preferable to it coasting off into the sunset.

To ensure that the camera is oriented in a horizontal position as it descends, we need to use two shock cords. The first will run from the nose cone, outside the payload body, and into the main rocket body, not entering the payload compartment. The second cord is attached to the transition piece of the payload compartment. Both cords are of equal length so that when hung underneath a parachute, the camera is in a near-horizontal position, illustrated in Figure 9-17. The shock cords are attached to the payload body at the opposite side to the lens. The parachute is tied to the shock cords loosely.

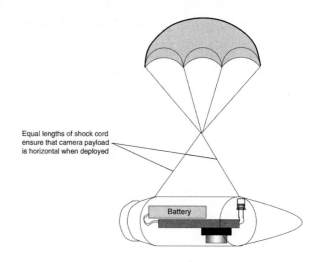

Equal lengths of shock cord ensure that camera payload is horizontal when deployed

Battery

Figure 9-17 *Model rocket camera shock cord detail.*

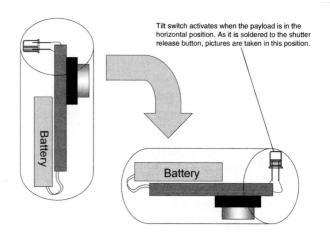

Tilt switch activates when the payload is in the horizontal position. As it is soldered to the shutter release button, pictures are taken in this position.

Battery

Battery

Figure 9-18 *Model rocket camera operation.*

It can be seen clearly in Figure 9-18 that, while the rocket is in the launch position, the tilt switch is in the open circuit position, and when the payload tilts over the tilt switch is in the closed position. It is therefore imperative at all times during pre-flight, from the moment the battery clip is connected, that the camera payload is kept in the vertical position. An assistant can prove very useful here to hold the payload while the rocket is prepared for launch.

Note: once the camera is switched on, it will deactivate within 30 s to conserve battery. As this is built into the firmware of the chip on board, there is little we can do to remedy this. As a result, you need to ensure that the launch sequence, take-off, and delay do not exceed this time. Try to stick to motors with a fairly short delay. Due to the extra weight that the camera payload will add to your model a short delay should not be a problem and has been shown to be successful.

Chapter 10

Model Rocket Movie Cameras

Even before the days of modern sophisticated electronics, model rocketeers were looking at ways of taking moving pictures from the sky.

The Estes Cineroc was the first model rocket kit to be commercially available; taking a special cartridge, it allowed cine film of the flight to be taken, which could be shown using a projector and a screen. The model was released in 1970, using the Super 8 film of the day to shoot at a brisk 31 frames per second. The result of the speed of the film, which was higher than the widely accepted 25 frames per second, was that the film appeared to be in "slow motion" when it was replayed.

More recently, cheap video CCD technology enables us to build video payloads at minimal cost. There are two methods presented in this chapter, and both are effective. The first method produces "live video," which is sent to a base station; the second records the video to be rendered into a moving picture on a PC once the images have been downloaded using a USB cable.

If building your own model rocket movie camera sounds like too much of a daunting task, Estes have released their "Oracle" model rocket kit (see

Figure 10-1   *Estes Oracle camera rocket kit.*

Figure 10–1), which will achieve the same result as the projects in this chapter, albeit without the satisfaction of knowing that you built it yourself.

Project 36: Build a Model Rocket Video Camera Payload (I)

You will need

Egglofter rocket

Micro-transmitting spy cam with transmitter and receiver

Polymorph

Tools

Junior hacksaw

Note

As the camera used in this project transmits on a frequency of 2.4 GHz, you should not need a license or other certification when operating this equipment in many countries. Check local legislation.

Camera technical specification – What does it all mean?

When looking at a camera, you will likely find a panel on the back of the box giving you all sorts of technical information. Here is a breakdown of what it all means, helping you understand a camera so that you can select a camera to meet your needs.

Camera specifications

Image sensor 1/3" CMOS CCD

This is the device used to record the image, it is analogous to the "negative" in a conventional camera. CMOS refers to the technology used in the camera; CCD stands for charge-coupled device, and is a term used to refer to the camera sensing element technology.

TV system: NTSC/PAL

This refers to the standard that the output from the receiver produces. NTSC is used in North America, PAL in the UK and much of Europe. Make sure that the camera and receiver you buy is suitable for use with the equipment you will be using to display the image. Be especially wary when buying equipment from online auction sites, to ensure that it complies with your local standard and to avoid disappointment.

Resolution: 380 TV lines

The resolution refers to the smallest detail that the image sensor can resolve. 380 lines is quite respectable.

Scan frequency: 60 Hz NTSC, 50 Hz PAL

The scan frequency refers to the number of times that the image is scanned. Hertz (HZ) means the number of times per second. For the NTSC system, the screen is scanned 60 times a second, for the PAL system, the screen is scanned 50 times a second. The eye recognizes images scanned at more than 25 times per second as continuous motion.

Minimum illumination: 1 lux

This specification refers to the minimum amount of light that the camera can work with. As you are likely to be launching your model rockets in the daytime, you should experience few problems in this respect.

Output power: 10 mW (UK)/3 mW (US)

The output power refers to the amount of power used to send the signal from transmitter to receiver; the strength of signal that can be used will largely depend on local legislation and regulations.

Transmission frequency: 2.4 GHz

This refers to the frequency at which the video signal is sent. In many countries 2.4 GHz is the frequency used for consumer devices without requiring a license.

Camera power: 5–9 V DC 200mA

This is the amount of power that the camera requires to operate. Thankfully, this can be supplied by a simple square PP3 9 V battery which will be sufficient for many, many, flights.

Receiver power: 9–12 V DC

This is the voltage rating for the receiver; as this is low voltage, it can be provided by a battery – ideal for use in a field!

Maximum transmitting distance: 328 ft/100 m

This is the furthest that the device will transmit the signal; bear this number in mind when sizing the motor you will use to power your flight, as ideally, you want to capture as much of the flight as possible. A little experimentation with your own camera setup is imperative, you will most likely find that you will achieve distances that far exceed this figure as there are no obstructions between the transmitter and the signal.

Operating temperature: 0–50°C/32–122°F

This is the temperature band that your camera and transmitter has been designed to work and operate at; it should seem commonsense that your camera should be isolated from any hot exhaust gases or recovery system gases that may cause incorrect operation of the device.

An egglofter rocket provides a good base for modification for this project; the rocket is designed to have a large weight at the front, and as a result, should remain stable when we load the egg compartment with our camera and transmitter.

In the past couple of years, with increasing miniaturization and the lowering of costs, small cameras intended for covert surveillance work have become readily available (see Figure 10-2). These generally operate on the 2.4 GHz band, which does not require a license if the power of the transmitters is modest. One of these cameras is eminently suitable for inclusion in a model rocket video camera payload.

Where can I get one of these spy cams?

www.hobbytron.com/
spyville.com/www.html
www.surveillance-spy-cameras.com/
mini-spy-camera.htm
www.raidentech.com/24ghzmiwicoc.html

Figure 10-2 *2.4 GHz covert surveillance camera.*

The battery for the camera is mounted inside the payload compartment of an egglofter rocket. The power lead runs outside to the main camera, which sits on the side of the rocket. While this is terrible from an aerodynamic point of view, it has proved to be successful. More advanced rocketeers could consider including the camera lens pointing out of the second stage of a rocket so that the first stage can be viewed peeling away from the model in flight. Certainly, this project provides many opportunities for diagnosing problems with multiple-stage rockets – seeing what is happening can be invaluable when trying to find out what went wrong.

For mounting, my favourite material Polymorph used in small quantities, can be a real boon here. We don't want the 9 V battery rattling about in the egglofter case, so make it a polymorph base and you will find it remains stable. Polymorph can also be used to make the support which holds the camera. See Chapter 3: The Model Rocketeer's Workshop for a recap of what Polymorph is and how to use it.

It is vital in this project to check the recovery system, and check again! The camera represents a significant investment and it would be undesirable to either damage it in a harsh landing, or even lose it as it flew into the distance and away over the treetops!

The camera has a short 5 cm wire aerial; be sure to have this sticking out of the model rocket in order to get the best reception (see Figure 10–3). Many cameras quote distances of around 100 m/350 ft; in practice, with model rocketry, you may well achieve far superior signal reception due to the fact that the model is going straight up into clear sky, and there will not be any buildings or objects between the path of the camera and the receiver! This situation may well change.

One of the things to consider when launching this rocket is how you are going to use the signal – are you simply going to watch the footage as it occurs, or record it for posterity? The chances are you will be out in the middle of a field launching rockets, so it is unlikely that there will be any mains power sources around. If you have a vehicle nearby, you could run a small video recorder from an inverter connected to your car's 12 V supply. If, however, you want to be completely mobile, a low-cost solution would be to use a laptop with a USB video capture device connected to the receiver.

You will be amazed when you look at the pictures captured from your model rocket's flight. Details are clearly discernible. The pictures in Figure 10-4 were originally taken in colour, but have been reproduced here in black and white. Figure 10-4i clearly shows the model rocket sitting on a standard rod launcher. The grass can be seen beneath the launcher – this particular launch site is a football pitch, and the grass, marked white, can be seen below.

Figure 10-3 *Camera mounted on the side of a commercially available egglofter kit. Image courtesy Peter Barratt.*

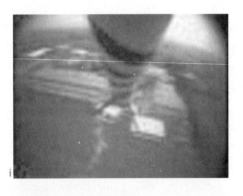

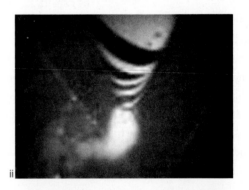

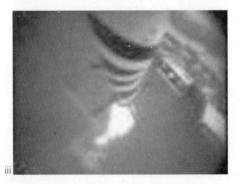

Figure 10-4 *Footage captured in flight.*

The button is pressed, and the rocket soars into the sky! In Figure 10-4ii we can see the plume of smoke and flame below the rocket. We can gauge the height of the model by looking at the lines on the ground; more of the football pitch is now clearly visible. Looking at Figure 10-4iii, we can see the rocket is high in the sky and we can just make out buildings in the background. In Figure 10-4iv we can see the rocket has clearly deviated from the vertical; the horizon can be seen, and large buildings in the background.

> If you are interested in developing this project further, there are a range of projects in 101 *Spy Gadgets for the Evil Genius,* including building your own video transmitters and receivers.

Project 37: Build a Model Rocket Video Camera Payload (II)

You will need

Completed model rocket camera payload

2 × normally closed reed switch

2 × strong magnet

Epoxy resin adhesive

In this project we will be using the camera's "continuous shooting" mode in order to capture a sequence of frames that can be later rendered into a video.

The crucial difference between this project and the last is the method of triggering the camera. Rather than the shutter firing after the recovery system has deployed, we want our camera to trigger right from the moment the rocket leaves the launch pad. We don't really want to trigger the camera before the rocket is launched because we only have a limited amount of memory in our Pencam, and as such do not want to waste precious shots.

In addition, it will be advantageous to add a mirror at 45° to the ground, allowing the side facing lens of our previously constructed camera project to face the ground. A front-surface mirror, where the reflective surface is on the front of the glass rather than the rear, is ideal for this. The position of the mirror is shown in Figure 10-5, the cowling has not been included in this diagram for clarity.

There are a variety of attachment methods; I have found that a combination of hot-melt glue and polymorph – as mentioned in Chapter 3 – work well, giving a solid mounting for the mirror. The flexibility afforded by the polymorph material in its fluid state allows it to be moulded into some semblance of an aerodynamic shape.

Returning to the shutter-release system, the method I have devised employs two normally closed reed switches. A reed switch comprises a glass envelope within which two fine metal contacts are enclosed.

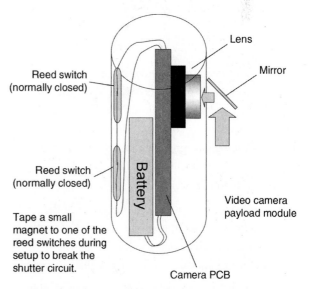

Figure 10-5   *Model rocket movie camera construction diagram.*

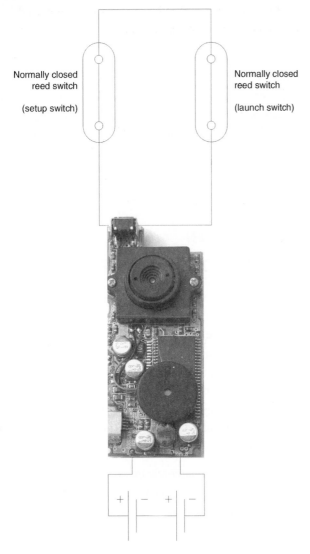

Normally closed
reed switch

(setup switch)

Normally closed
reed switch

(launch switch)

Figure 10-6 *Model rocket movie camera schematic.*

Whether the contacts are open or closed depends on whether a magnet is present or not. In this method, we will employ two "normally closed" reed switches. This is to say, the circuit is complete when NO magnet is applied. Holding a magnet close to the reed switch therefore keeps the circuit open. This is shown schematically in Figure 10-6.

The reed switches are glued to the inside of the payload tube. It is only thin plastic, and the magnetic field from an average magnet will be sufficient to penetrate the thin plastic and keep the contacts opened.

The reason that two reed switches are employed is that one switch serves to isolate the shutter circuit while

the rocket is set up; the other switch is used as the shutter release and can be tinkered with while the rocket is on the launch pad without wasting precious shots. The nice thing about reed switches is that they do not penetrate the rocket's envelope and as such do not present any aerodynamic resistance. If you only wanted to use one reed switch, you could always use a small unobtrusive push-to-break switch as the setup switch instead, although this would necessitate drilling a hole in the payload envelope, creating a little extra drag.

If you have decided to use a reed switch as a setup switch, you will need a small magnet to fix in place while you are setting up the rocket. This should be taped to the rocket body using a low-tack tape such as masking tape. When the rocket is ready to be "armed" for launch the magnet can be removed.

For the shutter trigger switch, a magnet can be attached to the end of a long rod or pea stick; when the rocket is set up on the launch pad, the rod is pushed into the ground next to the rocket, with the magnet resting on the rocket body next to the reed switch. This method requires no modification of the launch pad itself, and furthermore allows a good degree of flexibility of positioning the magnet.

The procedure for launching is as follows:

1. Set the rocket base on the stand.
2. Load the batteries into the camera payload module with a magnet taped to the side of the rocket over the setup switch or while holding in the setup switch.
3. Align the magnet on a stick next to the reed switch, ensuring that the camera display is not shooting pictures.
4. Ensuring that the trigger magnet is in place, remove the setup magnet or release the setup switch.
5. Press the camera "mode" switch to display "CT" for continuous shooting.
6. Retreat to a safe distance and launch within 30 s.

The PenCams are of an incredibly simple design; this simplicity is one of their strengths, especially for hackers and modifiers like us, although for this project, the simplicity of these cameras is also a weakness.

Many high-end digital cameras will quite happily export a video file that can be viewed instantly in Windows Media Player or other such application. Not so with the Aiptek PenCam, there is a little extra work to be done before we can view our rocket videos.

Take your movie payload with its precious cargo of space video, and connect it to your PC via the USB port. Using the software on the PenCam disk select "AVI Creator." This opens up an application that allows us to select multiple images and render them into a video. One of the nice things about this software is that you can select the frame rate. This way you can make slow-motion videos of your rocket's flight.

Taking it further

Model rocket flight launch videos are an excellent promotional tool for any model rocket club. Uploading a video of an actual rocket launch, filmed from the rocket's view to your club website is the height of cool, and will doubtless attract much attention to your club.

Chapter 11

Rocket Mail

If an evil genius wanted to send a message from his/her top secret lair somewhere deep in the forest, they would hardly call FedEx. Something far more devious and technically sophisticated would be employed…

Project 38: Rocket Mail

You will need

- Model rocket of your choice
- Brightly coloured paper
- Laminating film or sticky back plastic (optional)

Tools

- Pen
- Laminator (optional)

Here is a really fun project to try – launching a small payload in the form of an item of "mail." This can either be done to mark a special occasion or by launching a random message with contact details to see how far your payload carries. You might like to send a friend (or even yourself) a postcard on a special day such as a birthday; it is nice to have a piece of mail delivered by rocket! Alternatively, make up a label similar to that shown in Figure 11-1, replacing the dummy address with your own.

You need to make sure that when you prep the rocket before launch, the recovery wadding goes in before the message – this way, burning debris isn't ejected from your rocket at apogee! Check out Figure 11-2 for a diagrammatic view, and Figure 11-3 for a photo of how it is done.

When you insert the mail into the rocket, fold it sufficiently to ensure it fits in the tube, but do not fold too tightly. Ideally you want it to carry on the wind when the ejection system is deployed.

This message was launched by "Rocketmail" in an amateur model rocket experiment. I would be grateful if you can note the site where you found this message and contact me at:

gavindjharper@rocketmail.com

Gavin D. J. Harper
195 Rocket Street
Mars
Milky Way

1-CALL-071-1957

Figure 11-1   *Example message slip.*

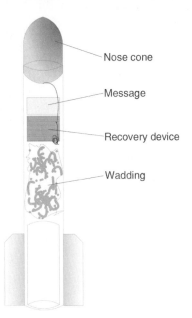

Figure 11-2 *Rocket mail diagram.*

Nose cone

Message

Recovery device

Wadding

Figure 11-3 *Prepping the rocket for launch.*

You might want to laminate the message to ensure that it does not get destroyed in wet weather. Lamination will of course increase weight, and probably decrease the distance the message will travel.

Caution

Do not try to "aim" your rocket in any way at anybody or anything.

Remember that you should never launch a rocket with the launch rod more than 30 degrees from the vertical.

Rocket mail history

If you feel a bit silly trying to send messages by rocket – don't worry, you're not the first, and you certainly won't be the last! Here's a little background information on the bold pioneers who went before!

Figure 11-4 *First day commemorative cover from the USS Barbero.*

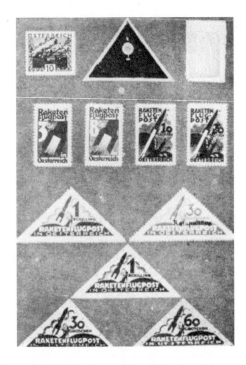

Figure 11-5 *Early Austrian rocket mail stamps. Image courtesy NASA.*

In 1931, a German by the name of Gerhard Zucker began experimenting on the problem of transporting mail by rocket. He demonstrated his idea to the Royal Mail in the UK on July 31 1934, but it was a failure. His rocket was 1.07 m long and 18 cm wide and on its maiden flight carried 1200 "first day covers." The rocket exploded on the launch pad, sending burning mail in all directions. By February 23 1936, the United States Postal Service (USPS) launched their first successful rocket mail from New Jersey to New York.

By 1959, the system had been perfected to the point where a cruise missile was launched from the USS Barbaro to a base in Florida successfully carrying two USPS mail containers. The commemorative first day cover pictured in Figure 11-4 was one of the items of mail on that flight. Rocket mail has never really taken off (pardon the pun), as it has been neither reliable nor cost effective.

There are people out there who collect stamps from rocket mail. Aerophilately is the name given to hobby of collecting airmail stamps, astrophilately therefore is the name given to people who collect rocket mail stamps. Figure 11-5 shows some early rocket mail stamps.

Chapter 12

Introduction to the Flight Computer

Over the years as modern electronics has progressed, model rocketry has grown in sophistication. Electronic devices have become smaller and smaller – the net result being that we can now make sophisticated payload electronics that fit within the tight size and weight constraints model that rocketry imposes.

The first Apollo Automatic Guidance Computer (AAGC) 4k words of RAM in magnetic core memory and 32k words of ROM in an outmoded technology called core rope memory. To demonstrate how truly *awesome* the projects you are about to construct really are, let's compare some of the details of the original Apollo Guidance Computer with our humble model rocket flight computer, and work out how humble it really is!

The original AAGC had 4k words (a word is 16 bits or two bytes in length) of RAM and 32k words of ROM, 36k words of memory in total. The "brain" of our flight computer, BASIC Stamp BS2p, has eight lots of 2k bytes EEPROM memory. In addition to this, we'll be using an external EEPROM memory for datalogging, with 32k bytes of storage space. In total, this is 48k bytes, or 24k words of memory. To put this in perspective, the flight computer you'll be building has two-thirds of the memory of the computer that landed astronauts on the moon!

The AAGC ran at a processor speed of 2MHz, while in comparison the Ubicom SX48AC processor at the heart of the BASIC Stamp that controls your model rocket flight computer runs at a blazing 20 MHz. This means that our model rocket flight computer runs 10 times faster than the Apollo Guidance Computer!

Building the flight computer

You will need a fairly sizable rocket to launch this payload. The payloader rocket in Chapter 5 comes highly recommended.

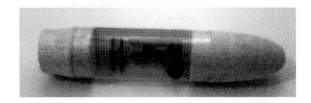

Figure 12-1   *The flight computer in the payload buy.*

I would recommend that you build this circuit on stripboard; this allows you to experiment, solder and de-solder at minimal cost, and with minimal development time.

I use strip sockets to make a prototyping area that I can stick my 8-pin chips in, this makes it fairly easy to swap between a range of chips.

Make sure that you use DIL sockets for the Stamp – these guys are really a little pricey to solder in permanently.

The flight computer is shown in the payload bay in Figure 12-1.

Datalogging

For storing the data that we record from various sensors, we have a couple of options. Parallax make a variant of the BASIC Stamp, the BS2pe, which has extra memory on board and is specifically designed for datalogging applications. However, the trade-off for this extra memory is program execution speed: the BS2pe runs at around 6000 instructions per second, compared to the BS2p's 12000 instructions per second.

This is also available in a variant the BS2p40, which has 32 input and output pins! This gives the hobbyist enormous potential for upgrade and expansion. The chip is shown in Figure 12-2.

The BS2p40 can be bought as part of a starter kit (see Figure 12-3), which includes a serial connection to

Figure 12-2   *The BS2p40 chip.*

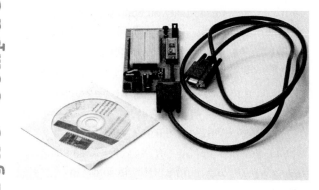

Figure 12-3   *The BS2p40 starter kit.*

```
        Tx  1        40
        Rx  2        39  GND
       ATN  3        38  RES
       GND  4        37  +5V
        P0  5        36  A15
        P1  6        35  A14
        P2  7        34  A13
        P3  8        33  A12
        P4  9        32  A11
        P5 10        31  A10
        P6 11        30  A9
        P7 12        29  A8
        P8 13        28  A7
        P9 14        27  A6
       P10 15        26  A5
       P11 16        25  A4
       P12 17        24  A3
       P13 18        23  A2
       P14 19        22  A1
       P15 20        21  A0
```

Figure 12-4   *The BS2p40 pinout.*
Image courtesy Parallax Inc.

your PC, all of the Parallax software that is needed to get things going, a few goodies and other components. This is the way to go for easy, hassle-free start up.

There are lots of good books out there that are a primer on how to use the Basic Stamp, the manual supplied with the development kit is quite clear and should get you up and running. For your reference, a pinout of the BS2p40 is reproduced in Figure 12-4.

The second option is to use some external memory. Microchip manufactures a nice range of Electronically Erasable Programmable Read-Only Memory (EEPROM) chips. For the flight computer, we will be using their 24LC256 EEPROM. This chip comes in an 8-pin DIL package, and can store 32,768 bytes of data. It uses an I2C interface, meaning that we only need to use two of our Stamp's precious I/O pins to read and write to it, and this is easily achieved using the BS2p's I2CIN and I2COUT commands. It is also possible to

chain together up to eight of these chips, and still access them using only two I/O pins, which could be very handy for projects requiring a lot of data storage space, such as bigger projects with lots of sensors or where we need to log data for a long time.

The Stamp is connected to the 24LC256 as shown in Figure 12-5.

To write data to the EEPROM we use the I2COUT command like this:

```
I2COUT pin, %10100000, address.HIGHBYTE\
address.LOWBYTE, [data]
```

Note that the EEPROM can only be connected to Stamp pins 0 and 1 or 8 and 9 (due to limitations of the Stamp), and so the pin must be either 0 or 8. The second argument (%10100000) is the control code to instruct the EEPROM to write data to the address in the EEPROM.

To read back our data, we use the I2CIN command:

```
I2CIN pin, %10100001, address.HIGHBYTE\
address.LOWBYTE, [data]
```

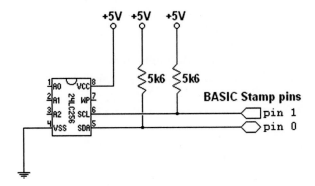

Figure 12-5 *Connection diagram for the Stamp to 24LC256.*

This time, the control code to read the information stored at address and store it in a variable called data is %10100001.

Lift off detection

We don't want to start collecting data from our rocket until it lifts off; otherwise, if the rocket sits on the launch pad for a long time, we'll end up filling our EEPROM with very uninteresting data from ground level. The most reliable method to detect launch is to use a dedicated acceleration switch. These switches close when they sense acceleration greater than their threshold level. They are commonly manufactured with thresholds of 2 g, 5 g, 10 g and 15 g. A 5g switch is recommended for use in the model rocket flight computer. This seems a good compromise; the switch must be sensitive enough to detect lift off, but not so sensitive that it closes too easily, and starts collecting data before launch.

The switch is connected to the Stamp according to Figure 12-6. When the switch is open (i.e. the rocket is not accelerating) then the input is pulled to ground, and is read by the Stamp as zero. When the rocket lifts off, the switch closes, the Stamp pin goes to +5V, and the Stamp detects this and begins to collect data.

Analog-to-digital converters

Some of the sensors we will be using output an analog voltage proportional to the parameter we are trying to

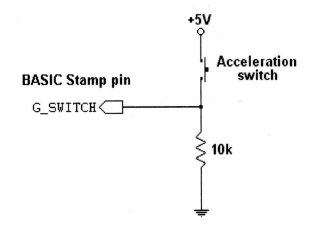

Figure 12-6 *Diagram showing operation of acceleration switch.*

measure. Unfortunately, the "brain" of our flight computer, the BASIC Stamp, cannot read analog signals directly. We need to convert this analog signal to a digital signal, which can be read and manipulated easily by the Stamp. For this we use an analog-to-digital converter, or ADC.

Maxim manufactures a wide range of ADCs, and the two we will be using in our flight computer are the MAX187 and MAX186. The MAX187 is a single-channel ADC, meaning it can read only one analog voltage. The ADC is connected to the Stamp according to Figure 12-7, and we communicate with it using a simple SPI serial interface as shown below:

```
LOW ADC_CS
SHIFTIN ADC_DATA, ADC_SCLK, MSBPOST,
[adc_value\12]
HIGH ADC_CS
```

The LOW ADC_CS command tells the ADC that we want it to send us a reading, the SHIFTIN command takes the reading into the Stamp, and HIGH ADC_CS tells the ADC that we are finished with it for the time being.

The MAX186 has eight analog inputs. To use the MAX186 ADC is slightly more complicated, but only slightly. This time, we need to use four Stamp pins to communicate with the ADC (see Figure 12-8); the extra one is because this time we need to send information to the ADC to tell it which analog channel we want to take a reading from. We do this using the SHIFTOUT command:

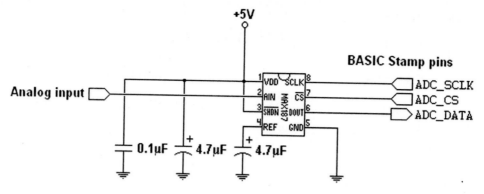

Figure 12-7 *MAX 187 connection diagram.*

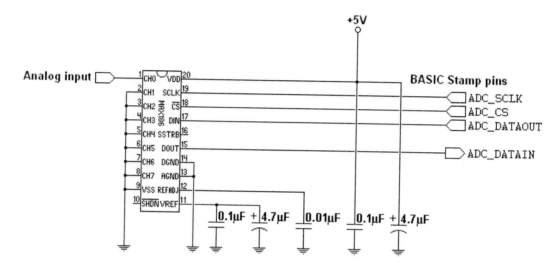

Figure 12-8 *MAX 186 connection diagram.*

LOW ADC_CS
SHIFTOUT ADC_DATAOUT, ADC_SCLK,
MSBFIRST, [channel]
SHIFTIN ADC_DATAIN, ADC_SCLK, MSBPOST,
[adc_value\12]
HIGH ADC_CS

Otherwise, communicating with the MAX186 is exactly the same as for the MAX187.

How to program your flight computer

To program the BASIC Stamp, the "brain" of your flight computer, you'll need the BASIC Stamp Editor software, available for free from Parallax. Hook your BASIC Stamp to your PC according to the instructions in the

BASIC Stamp Reference Manual (also available for free from Parallax). Then get the appropriate program for the project you've built from the "50 Model Rocket Projects for the Evil Genius" website, open it in the Editor, and upload it to the Stamp (how to do this is covered in the manual). The program runs each time that the flight computer is switched on, so install the flight computer in your rocket, switch it on, and begin the countdown!

All of the projects described in this chapter, with the exception of the first two, record data of some kind. When you have recovered your rocket, plug the flight computer back into your PC, open the appropriate data download program in the BASIC Stamp Editor, and upload it to the Stamp. The program will automatically begin to run and send the stored data back to your PC. This data can be analysed and plotted using spreadsheets available from the "50 Model Rocket Projects for the Evil Genius" website.

Some rocket projects may require triggering of recovery devices electronically, rather than using the ejection charge integral to the rocket motor. Some multi-stage rockets may need additional rocket motors to be ignited electronically, rather than simply igniting motors with quickmatch or by using the previous motor to ignite the next motor. In this instance, we can use our BASIC Stamp as a timer, getting it to trigger additional stages or recovery devices after a pre-set delay from lift-off.

This project uses an International Rectifier IRFD110 HEXFET (although many other FETs would be suitable) to discharge a 1000 μF capacitor through an electric match or flashbulb to fire our deployment

charge or ignite the next rocket motor stage. See the schematic in Figure 12-9 for how to wire the HEXFET to the Stamp. While the Stamp pin is LOW, the resistance between the source pin and drain pin of the HEXFET is very large. By making the output of the Stamp pin HIGH, the resistance drops to virtually nothing, allowing charge from the capacitor to flow through the electric match, igniting it.

The delay may be altered by simply changing the value of the constant DELAY in line 26 in the source (see CODE 1, Appendix 12-1), and downloading the updated program to the Stamp (the default value for the delay is 5 seconds).

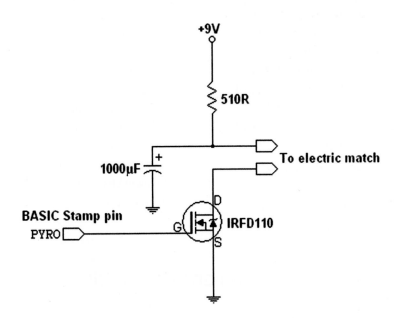

Figure 12-9 *HEXFET connection schematic.*

Project 40: Magnetic Apogee Sensor

It is possible to detect when a rocket tips over at the top of its flight path by monitoring a component of the Earth's magnetic field. As the rocket tilts horizontally, then a component of the field drops to zero. We can

monitor the field using a magnetoresistive sensor, and when it drops to zero, fire our deployment charge. This ensures that parachute deployment occurs as close as possible to apogee, exactly where we want it.

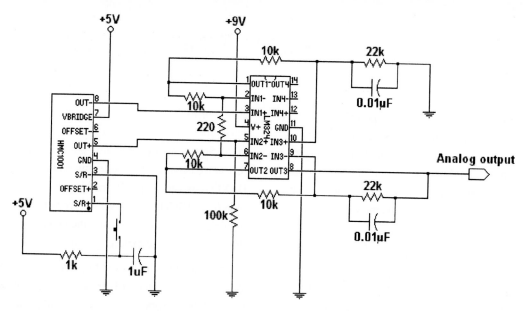

Figure 12-10   *HMC1001 connection schematic.*

The Earth's magnetic field is typically 0.5 Gauss, which is actually a very weak field, so our sensor needs to be pretty sensitive to detect this field. A couple of suitable sensors are the Phillips KMZ51 and the Honeywell HMC1001. The very small signal from the sensor needs to be amplified so that the analog-to-digital converter can read it. For this we use an operational amplifier, or op-amp.

The op-amp circuit in the schematic uses the HMC1001, although it would also work with the KMZ51 if the values of some of the resistors are altered. The op-amps take the output signal from the magnetoresistive sensor, and then offset and amplify it so that it is around +0.5 V when the rocket is aligned with the magnetic field, and around +4.0 V when the rocket is pointed straight up. The final analogue signal output from the op-amp is then fed into our analogue-to-digital converter. The Stamp reads the digitized signal from the ADC and monitors it for a when it reaches a minimum, and at that point fires the deployment charge. (see CODE 2, Appendix 12-1.) see Figure 12-10.

Project 41: Barometric Altimeter

This project records the air pressure throughout the rocket's flight. As the rocket's altitude increases, so the air pressure will decrease. We can measure this air pressure, and calculate how high our rocket has flown.

For the troposphere, the region of the atmosphere below 11,000 m (around 36,000 ft), air pressure, altitude and temperature are linked according to the following equations:

1. $P = 101.29 * [(T + 273.1) / 288.05] ^ 5.256$
2. $T = 15.04 - (0.00649 * h)$

Where P is pressure in kPa, T is temperature in Celsius, and h is altitude in m. Rearranging these equations gives us the equation for altitude from air pressure:

3. $h = - [44383.67 * (P / 101.29) ^ 0.19] - 44403.70$

From: http://www.grc.nasa.gov/WWW/K-12/airplane/atmosmet.html

Project 41: Barometric Altimeter

126

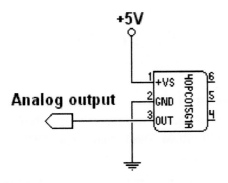

Figure 12-11 *The interface with the three supplied thermocouples.*

The pressure sensor we will be using is the Honeywell 40PC015G1A. This sensor can measure from 0 to 15 psi, or in metric terms from vaccum to slightly more than one atmosphere. The analogue output is a voltage from + 0.5 to + 4.5 V, which is proportional to pressure. This analog voltage is fed into a MAX187 ADC, converted to a digital signal, and sent to the Stamp. Our flight computer stores this in its EEPROM using the program

below, and then when we recover the rocket after its flight, we can download the data to a PC and calculate air pressure, and from that, our rocket's altitude (see CODE 3, Appendix 12-1.)

The sensor's analog voltage output is fed into a Maxim MAX187 analog-to-digital converter, and our BASIC Stamp controller will read the digital output and store it in our 24LC256 external EEPROM.

The circuit diagram for the whole shebang is shown in Figure 12-12.

After the flight, connect the flight computer to your PC, and download the program "DataDownload1.BAS" to the Stamp (see CODE 4, Appendix 12-1). The Stamp will then send the data back to your PC. You can then simply copy and paste the data to the Excel spreadsheet "BarometricAltimeter.XLS" that you will find on the "50 Model Rocket Projects for the Evil Genius" website, and it will calculate and plot the rocket's altitude during its flight.

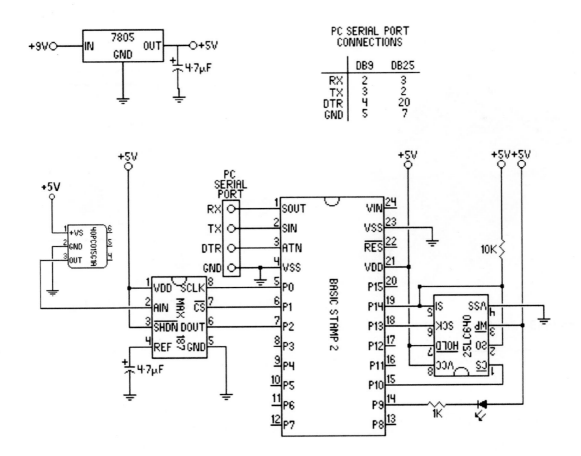

Figure 12-12 *Barometric altimeter circuit schematic.*

Project 42: Temperature and Humidity

This project allows you to take temperature and humidity readings of the atmosphere. Try flying your rocket in different weather conditions to sample a range of temperatures and humidities. If the cloud base is low enough, you may be able to sample a cloud. But try not to lose your rocket, and make sure never to fly in a thunderstorm!

Sensirion manufacture a sensor called the SHT11, which measures temperature and humidity. With the SHT11 Sensor Module, Parallax have mounted the sensor in a convenient 8-pin DIL package. This is shown in Figure 12-13. This module is very easy to work with: it uses a simple two-wire serial interface, and requires only four connections, +5 V, ground, and the serial clock and data lines to the Stamp.

When lift off is detected, the flight computer begins to take sensor readings and stores them in the external EEPROM. This is done using CODE 5 (see Appendix 12-1)

After recovering your rocket, because two channels of data have been recorded (temperature and humidity), we need to use the "DownloadData2.BAS" program. This is listed in Appendix 12-1 (CODE 6) and is also available from the website. The procedure is exactly the same: plug your flight computer into your PC, and download the program to the Stamp, and it'll send both channels of data back to your computer. Get the "TempHumidity.xls" spreadsheet from the "50 Model Rocket Projects for the Evil Genius" website, copy and paste your data into it, and it'll plot air temperature and humidity for the duration of your rocket's flight.

Figure 12-13   *The Sensiron SHT 11 sensor module.*

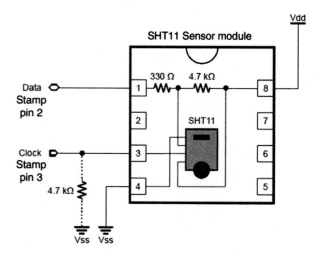

Figure 12-14   *SHT 11 connection diagram.*

Project 43: Acceleration

This project uses an accelerometer, a sensor that measures acceleration. From the rocket's acceleration, its velocity and the distance it has traveled can be calculated. You could try changing the nose cone profile of your rocket to see how this affects its speed. Or experiment with the fins: how does the shape, thickness and number of fins affect performance?

Analog Devices manufacture a neat range of accelerometers. The one we will use for this project is the ADXL78. This is available in three types, each with a different range of acceleration that they can measure: ±35 g, ±50 g, or ±70 g. The model that measures up to ±35 g, the AD22279-A, will be sufficient for our model rocket. This sensor outputs an analog voltage that is proportional to acceleration, similar to the pressure sensor in Project 40.

Sensor details

The sensor consists of a series of micromachined fixed plates and moving plates. These plates form a capacitor. When the sensor undergoes an acceleration, the plates move and the capacitance changes. This change in capacitance is measured using some electronic wizardry on board the chip, and an analog voltage proportional to the acceleration is output.

Unfortunately, Analog Devices only manufacture this sensor in a very small chip that has no leads, which means it is a real pain to solder. Luckily, we only need to make three connections to the chip; +5 V, ground, and analog out, and using a soldering iron with a small bit, and some steady hands, it is possible (see Figure 12-15)

The ADXL78 sensor measures acceleration in one direction, which runs along the axis of the chip from pin 4 to pin 8 (check the datasheet from Microchip if you are unsure which pin is which). We need to ensure that this axis is aligned with the direction of flight of our rocket. The listing for the program is CODE 7 in Appendix 12-1.

This project works in exactly the same way as the barometric altimeter in the previous project, except, of course, that the pressure sensor is replaced by the accelerometer. As in the barometric altimeter project, the acquired data is downloaded to the PC using the "DownloadData1.BAS" program. On the "50 Model Rocket Projects for the Evil Genius" website, you will find an Excel spreadsheet named "Acceleration.xls" that takes the downloaded raw acceleration data and integrates it against time to calculate speed and altitude.

Project 44: Vibration (2-Axis Accelerometer)

When a rocket lifts off, it experiences some serious vibration from the air buffeting it as it accelerates – just ask any Apollo or Space Shuttle astronaut! This project is designed to measure the vibration that our model rocket experiences during its flight.

The sensor we are going to use is another of Analog Devices' accelerometers, this time the ADXL320. The ADXL320 comes in a very small (4 mm × 4 mm × 1.75 mm) package that has no leads, making it very difficult to solder by hand. Help is at hand, however, as Analog Devices produce an evaluation board for their

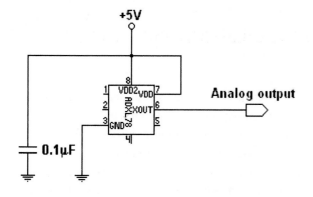

Figure 12-15 *ADXL 78 connection diagram.*

sensor (part number ADXL320EB), which includes a 5-pin header with a standard 0.1″ pitch to connect to your circuit.

This sensor measures ±5 g of acceleration in two axes. We want to measure vibration at right angles to our rocket's flight path, so once again we'll need to make sure that our sensor is orientated correctly. The

sensor measures vibration in the plane of the evaluation board, so the board needs to be mounted at right angles to the flight computer board.

This time, because we are measuring two channels of data, we need an analog-to-digital converter that can measure more than one analog voltage, such as the Maxim MAX186 ADC. This chip is capable of measuring up to eight channels. Although we'll only use two of them in this project, we'll use the other channels in a later project. Figure 12-17 shows how the ADXL320 is connected to the MAX186 8-channel ADC.

Like in Project 42, where we recorded air temperature and humidity, we have two channels of data to plot, so we must use the "DownloadData2.BAS" program. Copy and paste this data to the "Vibration.xls" spreadsheet, which is available from the "50 Model Rocket Projects for the Evil Genius" website, and it'll show you just how much your rocket vibrates during its flight (see CODE 8, Appendix 12-1).

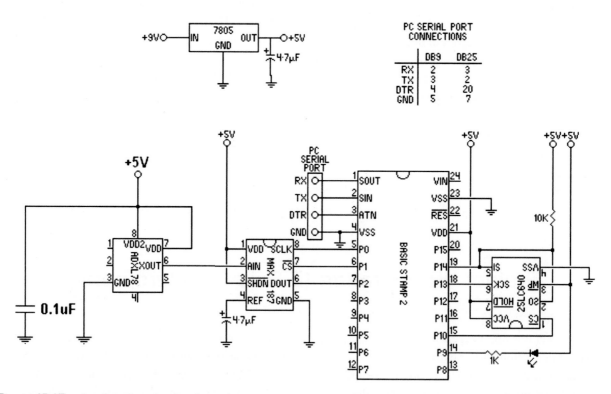

Figure 12-16 *Acceleration circuit schematic.*

You may have noticed that many "real" rockets are painted with black and white bands around them (for example, check out a video of a Saturn V launch). These bands enable rocket engineers and scientists to gauge how fast the rocket is spinning about its axis.

Unfortunately, our rockets take off so fast, and are so small compared to NASA-sized rockets that we need to look at alternative methods of tracking roll rate; they are simply too small for us to gauge visually how fast they are spinning. We need a fixed point of reference. The sun provides a very good point of reference. In the few seconds our rocket is in the air, the sun stays relatively fixed in the sky. If we mount an optical sensor on our rocket, we can look at the light levels as our rocket spins and then extract this data to our PC at base camp. By looking at the data, we can see the points when our rocket was looking at the sun. This is illustrated in Figure 12-19.

By adding an optical sensor, it is possible to determine how fast our rocket is spinning. The proper technical term is "roll rate."

The Stamp is very good at working with pulses, counting them and measuring frequency. Rather than having to go through the rigmarole of using an analog-to-digital converter or another unnecessarily complex solution, we can instead use a chip that outputs a frequency that is proportional to the amount of incident light.

For this project we will be using the TSL230, shown in Figure 12-20, a light sensor chip from Texas Instruments that is very easy to interface to the BASIC

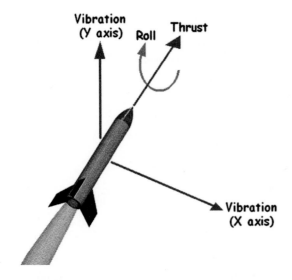

Figure 12-17 *The forces acting on a rocket in flight.*

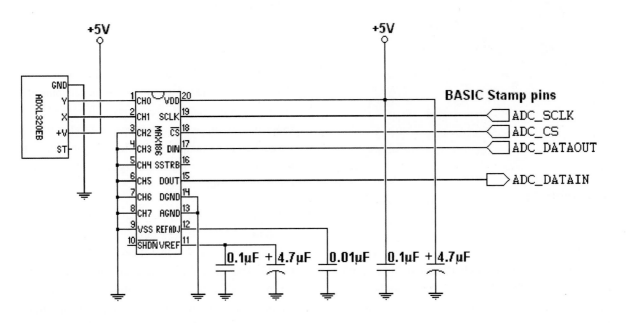

Figure 12-18 *ADXL320 connection diagram.*

131

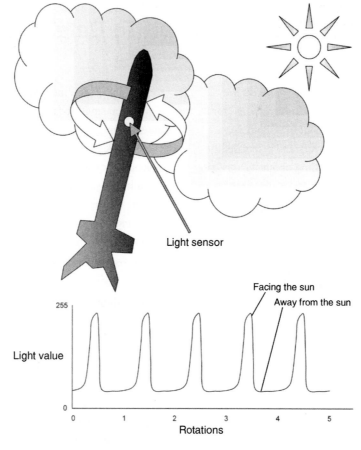

Figure 12-19   *How light level relates to roll rate.*

Figure 12-20   *Texas Instruments TSL230 sensor.*

Stamp (Pin 6 on the TSL320 simply connects to I/02 (Pin 7) of the BASIC Stamp II.) The TSL230 is quite a sophisticated sensor. Rather than just a single light-sensing element, it is in fact a scalable array of photo sensing elements. The chip allows us to switch on or switch off parts of the sensor, depending on the sensitivity that we require. Depending on the output required, all or part of the array can be used.

The output of the TSL230 is a square wave whose frequency is directly proportional to the intensity of light falling on the sensor. So as the rocket spins and the sensor faces the sun, the frequency will peak. The frequency of the peaks will correspond to the roll rate of the rocket. For example, if the frequency peaks 12 times per second, then we know our rocket is also spinning 12 times per second.

In bright light, the output frequency of the sensor can be up to 1.4 MHz. If we examine the capabilities of our BASIC Stamp, we find that it can accept frequencies up to 125 kHz. Thankfully, help is at hand, as our light sensor allows us to divide its output by 2, 10 or 100 if required. Dividing by 100 will give us a happy figure of 14 kHz as our maximum frequency.

The software to log the data is Rollrate BAS listed in Appendix 12-1 (CODE 9) and also available from the "50 Model Rocket Projects for the Evil Genius" website.

The sensor needs to be aligned so that it points horizontally out of the side of the rocket – don't forget to put a hole or window in the payload bay so that the sensor actually sees some light! Download the data to your PC using the "DataDownload1.BAS" program, and analyse it using the "RollRate.XLS" spreadsheet from the "50 Model Rocket Projects for the Evil Genius" website. The spreadsheet will plot light levels during the flight, and tell you how fast your rocket was spinning.

Try experimenting with your rocket to make it spin. Perhaps try adding angled tabs to the trailing edge of the fins to see how the roll rate is affected. Even with a rocket not designed to spin, slight imperfections in mounting the fins can cause it to spin.

Gyroscope measuring

An alternative method for measuring roll rate is using a gyroscope. Analog Devices makes a range of relatively inexpensive gyroscopes that come on a chip, and output an analog voltage proportional to the roll rate. The most suitable for our purposes is the ADXRS300. An evaluation board is available for this gyro, which comes in a DIL package complete with decoupling capacitors, that makes it very easy to use. Note however that the maximum roll rate that is measurable using this sensor is ± 300°/second, which makes it unsuitable for rockets that spin very quickly. If your rocket has canted fins designed to make it spin, the ADXRS300 may not be able to keep up! This sensor can be interfaced to the BASIC Stamp in the usual way, using an analog-to-digital converter.

Project 46: Exhaust Temperature (Thermocouple)

With this project, we will be adding a temperature sensor to our model rocket flight computer to monitor the temperature of the exhaust gases of our model rocket engine. There are lots of different experiments that we can do as a result of this. Do different brands of rocket motors burn any hotter? How hot is the exhaust during the "coast" phase compared to the "thrust" phase?

Because our sensor will encounter very high temperatures, many sensors are inappropriate. What we need is a sensor that is reasonably sensitive over a wide range of temperatures, but that can also withstand the heat of the rocket exhaust. For this we'll be using a thermocouple.

Parallax produce a thermocouple kit (see Figure 12-21) which uses a DS2760 lithium battery monitor to read the very small Seebeck voltages

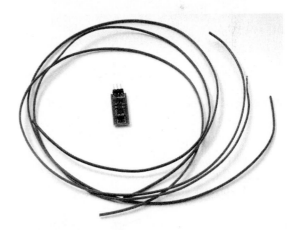

Figure 12-21 *Parallax thermocouple kit.*

output by the thermocouple. This module is easily interfaced to the BASIC Stamp BS2p via a One-Wire interface.

Seebeck voltage

In 1821, Thomas Johann Seebeck, an Estonian physicist, discovered that when a conductor was subject to a thermal gradient, it would generate a small voltage, now known as the Seebeck voltage. Thermocouples exploit this effect by providing an output voltage that is dependent upon:

1. the difference in temperature between two points
2. the metals used in the thermocouple construction.

The thermocouple consists of two dissimilar metal wires twisted together. One end of the wires is heated, and the other remains cool. It is the temperature gradient between the hot end (the hot junction) and the cool end (the cold junction) that produces the Seebeck voltage. This voltage is very small, usually on the order of a few mV.

Prepare the thermocouple hardware and wire it to the Stamp according to the instructions included with the thermocouple kit (see Figures 12-22 to 12-25).

The thermocouple needs to be threaded through the body of the model rocket – a little material will need to be removed from the motor mount so that the thermocouple can protrude into the exhaust and take measurements from the exhaust gases. Note that only two of the thermocouples supplied in the kit are suitable for this project, the Type K (Chrome/Alumel) and Type J (Iron/Constantan) thermocouples. The third thermocouple, the Type T (Copper/Constantan), is only rated to 400°C.

Upload the "Thermocouple.BAS" program listed in Appendix 12-1 (CODE 10) to your Stamp. During

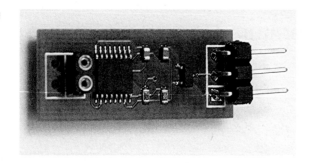

Figure 12-22 *Thermocouple interface board.*

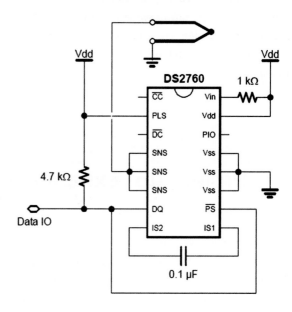

Figure 12-24 *Thermocouple schematic.*

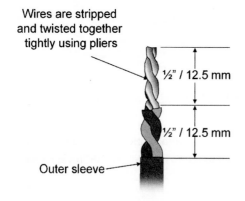

Figure 12-23 *How to prepare the thermocouple.*

Wires are stripped and twisted together tightly using pliers

½" / 12.5 mm

½" / 12.5 mm

Outer sleeve

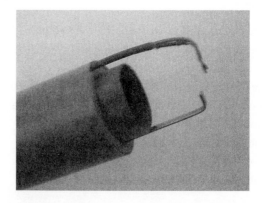

Figure 12-25 *Mounting the thermocouple.*

flight, the flight computer saves the raw data from both the hot and cold junctions of the thermocouple in the external EEPROM as usual. The "DownloadDataTC.BAS" (CODE 11, Appendix 12-1) program is used to calculate the hot junction temperature and upload it to your PC. This program uses a lookup table to calculate the hot junction temperature; you'll notice that you need to tell your flight computer which thermocouple you used, so that it uses the correct lookup table. The temperature data can be plotted using the "Thermocouple.XLS" spreadsheet.

Lookup tables for the "J" and "K" type thermocouples are given in Appendix 12-2.

Project 47: Full Data Acquisition System

This project is going to combine several of the preceding projects, to build a complete data acquisition system for our rocket. We'll be using five analog sensors, measuring air pressure, acceleration, vibration, roll rate, and magnetic field strength, interfaced to the BASIC Stamp via the 8-channel MAX186 ADC, and also the SHT11 temperature and humidity sensor interfaced directly to the Stamp. The Stamp will read each sensor in turn, and store the data in EEPROM. Figure 12-26 shows a schematic.

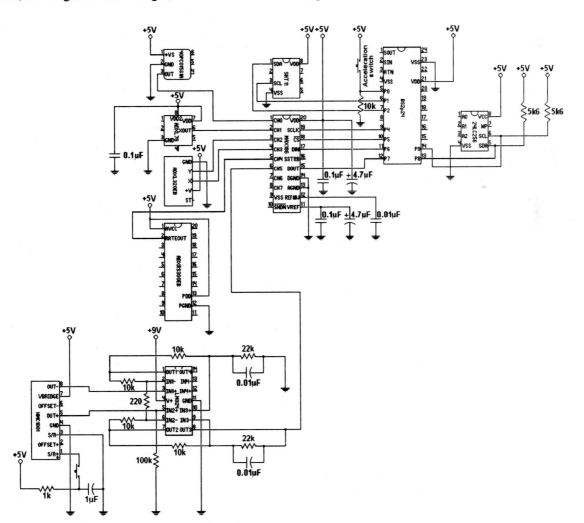

Figure 12-26 *Schematic for fill data acquisition system.*

For measuring roll rate, this time we recommend using the Analog Devices ADXRS300 gyroscope sensor that was mentioned in Project 43, as most of our other sensors are interfaced to the Stamp through the ADC, although there is no reason why the light sensor could not be used with some code rewriting.

This is one project where you will really need the extra speed of the BS2p as opposed to one of the slower BASIC Stamp variants. Try to read from so many sensors using a slower Stamp and you may only get a couple of readings per second from each sensor.

You may also want to add some extra external EEPROM memory to your flight computer so that you can store more data from longer flights. This is easily achieved by daisy-chaining together 24LC256 chips – up to eight can be chained in this way. The schematic showing how this is done is shown in Fig. 12.26. To read or write to the extra chips you'll need a slightly different control code. The control code should be changed to %1010aaa0 to write to the EEPROM, and %1010aaa1 to read from it, where aaa corresponds to the state of the address pins A2, A1 and A0 on the chip you want to access. For example, to write to the leftmost EEPROM in the schematic diagram, the control code you'd use is %10100110. Note that you only need to pull the address pins to +5V to set the bit; there's no need to pull them low as they are internally tied to ground. For more details, check out the datasheet, available from Microchip's website.

You may also want to add a capability to your computer to deploy recovery systems at apogee; integrating the code from Project 39 to do this into the code for this project is left as an exercise for the reader.

This time, you'll need to use the "DownloadData8.BAS" program to upload all eight channels of logged data from the Stamp to your PC. This data can be copied-and-pasted into the "DataAcquisitionSystem.xls" spreadsheet. This will give you full data on your rocket's flight, and plot various graphs of the data.

See CODE 12 and CODE 13 in Appendix 12-1 for source codes.

Project 47: Full Data Acquisition System

136

Appendix 12-1. Source Codes for BASIC Stamp Program

CODE 1

```
==============================================================================
''    Project 39
'     DelayTimer.bas
'     {$STAMP BS2p}
'     {$PBASIC 2.5}

==============================================================================
' -----[ Program Description ]-------------------------------------------------
' This program waits for lift off, waits for a set number of seconds, and then
'   fires the pyro charge. TO alter the delay (default is 5 seconds)
' change the value of the constant DELAY ON line 26.
' ----------------------------------------------------------------------------

' -----[ Pin assignments ]-----------------------------------------------------
PYRO         PIN        0          ' Pyro charge
G_SWITCH     PIN        1          ' Acceleration switch
' -----[ Constant definitions ]------------------------------------------------
DELAY        CON        5          ' delay in seconds
' -----[ Initialisation ]------------------------------------------------------
LOW PYRO                           ' Ensures that pyro is turned off
INPUT G_SWITCH                     ' Sets g-switch pin to input
' -----[ Main program ]--------------------------------------------------------
wait_for_liftoff:
   IF G_SWITCH=0 THEN wait_for_liftoff     ' Waiting for lift off to be detected
PAUSE DELAY * 1000                 ' Because PAUSE uses milliseconds
HIGH PYRO                          ' Fire pyro charge
END
```

```
===============================================================================
''    Project 40
'     MagneticApogeeDetection.bas
'     {$STAMP BS2p}
'     {$PBASIC 2.5}

'===============================================================================
' -----[ Program Description ]-------------------------------------------------
'' This program reads magnetic field strength from the ADC, and searches for
'  a minimum point in field strength. It then fires the deployment charge.
'  8 readings are taken and totalled, and compared to the total of the
'  previous 8 readings to prevent premature deployment due to noise.
'  -----------------------------------------------------------------------------

' -----[ Pin assignments ]-----------------------------------------------------
PYRO        PIN       0          ' Pyro charge
G_SWITCH    PIN       1          ' Acceleration switch

ADC_SCLK    PIN       0          ' ADC serial clock pin
ADC_CS      PIN       1          ' ADC chip select pin
ADC_DATA    PIN       2          ' ADC serial data pin

' -----[ Variable definitions ]------------------------------------------------
adc_value VAR         Word       ' Value read from ADC

total       VAR       Word       ' Current total of ADC readings

prev_total VAR        Word       ' Previous total of ADC values
i           VAR       Byte       ' Loop counter

' -----[ Initialisation ]------------------------------------------------------
LOW PYRO                         'Ensures that pyro is turned off

INPUT G_SWITCH                   'Sets g-switch pin to input

HIGH ADC_CS                      'Switches off ADC

total = 0                        'Ensures that total < prev_total for first

prev_total = 65535               'cycle through do...loop

' -----[ Main program ]--------------------------------------------------------
Wait_for_liftoff:

  IF G_SWITCH=0 THEN wait_for_liftoff    ' Waiting for lift off to be detected
```

```
Flight:
  total = 0
  FOR i=1 TO 8
    GOSUB Read_ADC
    total =  total + adc_value
  NEXT
  IF total > prev_total THEN Apogee_detected
  prev_total = total
GOTO Flight
Apogee_detected:
  HIGH PYRO                         ' Fire deployment charge
END
' -----[ Subroutines ]-------------------------------------------------
Read_ADC:
  LOW ADC_CS
  SHIFTIN ADC_DATA,ADC_SCLK,MSBPOST,[adc_value\12]
  HIGH ADC_CS
RETURN
```

CODE 3

```
'==============================================================================
'     Project 41
'     BarometricAltimeter.bas
'     {$STAMP BS2p}
'     {$PBASIC 2.5}

'==============================================================================
' -----[ Program Description ]-------------------------------------------------
' This program reads air pressure data from an ADC and stores it in
' external EEPROM until the EEPROM is full.
' ----------------------------------------------------------------------------

' -----[ Pin assignments ]-----------------------------------------------------
G_SWITCH        PIN     1           ' Acceleration switch

ADC_SCLK        PIN     0           ' ADC serial clock pin
ADC_CS          PIN     1           ' ADC chip select pin
ADC_DATA        PIN     2           ' ADC serial data pin

EEPROM          PIN     0           ' EEPROM data pin is 0; EEPROM clock pin is 1

' -----[ Constant definitions ]------------------------------------------------
EEPROM_WRITE CON    %10100000       ' Control code to perform write operation to
                                      I2C EEPROM

EEPROM_SIZE  CON    32768           ' Size in bytes of EEPROM
' -----[ Variable definitions ]------------------------------------------------
adc_value VAR       Word            ' Value read from ADC

address   VAR       Word            ' Address in EEPROM to write data

eeprom_data  VAR    Byte            ' Data byte to be written to EEPROM
' -----[ Initialisation ]------------------------------------------------------
INPUT G_SWITCH                      ' Sets g-switch pin to input
HIGH ADC_CS                         ' Switches off ADC

address = 0
' -----[ Main program ]--------------------------------------------------------
Wait_for_liftoff:
  IF G_SWITCH=0 THEN wait_for_liftoff    ' Waiting for lift off to be detected
DO

  GOSUB Read_ADC

eeprom_data = adc_value.HIGHBYTE
```

```
    GOSUB Write_EEPROM

    eeprom_data = adc_value.LOWBYTE
    GOSUB Write_EEPROM

LOOP WHILE address < EEPROM_SIZE

END

' -----[ Subroutines ]------------------------------------------------

Read_ADC:
    LOW ADC_CS
    SHIFTIN ADC_DATA,ADC_SCLK,MSBPOST,[adc_value\12]
    HIGH ADC_CS
RETURN

Write_EEPROM:
    I2COUT EEPROM, EEPROM_WRITE, address.HIGHBYTE\address.LOWBYTE, [eeprom_data]
    address = address + 1
RETURN
```

```
==============================================================================
'    DataDownload1.bas
'    {$STAMP BS2p}
'    {$PBASIC 2.5}

==============================================================================
' -----[ Program Description ]-------------------------------------------
' This program...
'
' -----------------------------------------------------------------------
' -----[ Pin assignments ]-----------------------------------------------
EEPROM          PIN    0          ' EEPROM data pin is 0; EEPROM clock pin is 1
' -----[ Constant definitions ]------------------------------------------
EEPROM_READ    CON    %10100001 ' Control code to perform read operation to
                                   I2C EEPROM
EEPROM_SIZE    CON    32768      ' Size in bytes of EEPROM
' -----[ Variable definitions ]------------------------------------------
address         VAR    Word       ' Address in EEPROM to write data
eeprom_data    VAR    Word        ' Data byte to be written to EEPROM
' -----[ Initialisation ]------------------------------------------------
address = 0
' -----[ Main program ]--------------------------------------------------
DO
  I2CIN EEPROM, EEPROM_READ, address.HIGHBYTE\address.LOWBYTE,
[eeprom_data.HIGHBYTE]
  address = address + 1

  I2CIN EEPROM, EEPROM_READ, address.HIGHBYTE\address.LOWBYTE,
[eeprom_data.LOWBYTE]
  address = address + 1

  DEBUG DEC eeprom_data, CR
LOOP WHILE address < EEPROM_SIZE
END
```

```
=================================================================
'    Project 42
'    TempHumidity.bas
'    {$STAMP BS2p}
'    {$PBASIC 2.5}

=================================================================
' -----[ Program Description ]------------------------------------
'' This program reads temperature and humidity data from the SHT11 sensor,
' and stores it in external EEPROM until the EEPROM is full.
' -----------------------------------------------------------------
' -----[ Pin assignments ]----------------------------------------
G_SWITCH        PIN     1       ' Acceleration switch
TH_DATA         PIN     2       ' Sensor data pin
TH_CLOCK        PIN     3       ' Sensor clock pin
EEPROM          PIN     0       ' EEPROM data pin is 0; EEPROM clock pin is 1
' -----[ Constant definitions ]-----------------------------------
TEMP            CON     %00011      'Control codes to read temperature and
HUMID           CON     %00101      'humidity from SHT11 sensor
STATUS_WRITE    CON     %00110      'Control codes to read and write SHT11
STATUS_READ     CON     %00111      'status register
EEPROM_WRITE    CON     %10100000   'Control code to perform write operation to
                                    ' I2C EEPROM
EEPROM_SIZE     CON     32768       ' Size in bytes of EEPROM
' -----[ Variable definitions ]-----------------------------------
sensor_data     VAR     Byte        ' Data read from sensor
ctrl_byte       VAR     Byte        ' Control byte sent to sensor
ack_bit         VAR     Bit
time_out        VAR     Bit
address         VAR     Word        ' Address in EEPROM to write data
eeprom_data     VAR     Byte        ' Data byte to be written to EEPROM
temperature     VAR     Word
humidity        VAR     Word
i               VAR     Byte        ' Loop counter
' -----[ Initialisation ]-----------------------------------------
INPUT G_SWITCH                      ' Sets g-switch pin to input
```

```
SHIFTOUT TH_DATA, TH_CLOCK, LSBFIRST, [$FFF\9]
GOSUB TH_Start_Sequence          ' Resets connection to temp/humidity sensor
address = 0
' -----[ Main program ]-------------------------------------------------------
Wait_for_liftoff:
  IF G_SWITCH=0 THEN wait_for_liftoff    ' Waiting for lift off to be detected
  DO
    GOSUB Read_Temp
    eeprom_data = temperature.HIGHBYTE
    GOSUB Write_EEPROM
    eeprom_data = temperature.LOWBYTE
    GOSUB Write_EEPROM
    address = address + 1
    GOSUB Read_Humidity
    eeprom_data = humidity.HIGHBYTE
    GOSUB Write_EEPROM
    eeprom_data = humidity.LOWBYTE
    GOSUB Write_EEPROM
    address = address + 1
  LOOP WHILE address < EEPROM_SIZE
  END
' -----[ Subroutines ]--------------------------------------------------------
Read_Temp:
  GOSUB TH_Start_Sequence
  ctrl_byte = TEMP
  GOSUB Send_Ctrl_Byte
  GOSUB Wait_For_Ack
  ack_bit = 0
  GOSUB Read_Data
  temperature.HIGHBYTE = sensor_data
  ack_bit = 1
  GOSUB Read_Data
  temperature.LOWBYTE = sensor_data
RETURN
Read_Humidity:
  GOSUB TH_Start_Sequence
  ctrl_byte = HUMID
  GOSUB Send_Ctrl_Byte
```

```
      GOSUB Wait_For_Ack
      ack_bit = 0
      GOSUB Read_Data
      humidity.HIGHBYTE = sensor_data

      ack_bit = 1
      GOSUB Read_Data
      humidity.LOWBYTE = sensor_data
RETURN

TH_Start_Sequence:
      INPUT TH_DATA
      LOW TH_CLOCK
      HIGH TH_CLOCK
      LOW TH_DATA
      LOW TH_CLOCK
      HIGH TH_CLOCK
      INPUT TH_DATA
      LOW TH_CLOCK
RETURN

Read_Data:
      SHIFTIN  TH_DATA, TH_CLOCK, MSBPRE, [sensor_data]      ' get byte
      SHIFTOUT TH_DATA, TH_CLOCK, LSBFIRST, [ack_bit\1] ' send ack bit
      INPUT TH_DATA                                     ' release data line
RETURN

Wait_For_Ack:
      INPUT TH_DATA                                ' data line is input
      FOR i = 1 TO 250                       ' give ~1/4 second to finish
        time_out = INS.LOWBIT(TH_DATA)            ' scan data line
        IF (time_out = 0) THEN EXIT        ' if low, we're done
        PAUSE 1
      NEXT
RETURN

Send_Ctrl_Byte:
      SHIFTOUT TH_DATA, TH_CLOCK, MSBFIRST, [ctrl_byte]    ' send byte
      SHIFTIN  TH_DATA, TH_CLOCK, LSBPRE, [ack_bit\1]    ' get ack bit
RETURN

Write_EEPROM:
      I2COUT EEPROM, EEPROM_WRITE, address.HIGHBYTE\address.LOWBYTE, [eeprom_data]
RETURN
```

Source Codes for BASIC Stamp Program

```
===============================================================================
'     DataDownload2.bas
'     {$STAMP BS2p}
'     {$PBASIC 2.5}

===============================================================================
' -----[ Program Description ]--------------------------------------------------
' This program...
' -----------------------------------------------------------------------------
' -----[ Pin assignments ]-----------------------------------------------------
EEPROM        PIN    0               'EEPROM data pin is 0; EEPROM clock pin is 1
' -----[ Constant definitions ]------------------------------------------------
EEPROM_READ   CON    %10100001       ' Control code to perform read operation to
                                     ' I2C EEPROM
EEPROM_SIZE   CON    32768           ' Size in bytes of EEPROM
' -----[ Variable definitions ]------------------------------------------------
address       VAR    Word            ' Address in EEPROM to write data
eeprom_data   VAR    Word            ' Data byte to be written to EEPROM
' -----[ Initialisation ]------------------------------------------------------
address = 0
' -----[ Main program ]--------------------------------------------------------
DO
  I2CIN EEPROM, EEPROM_READ, address.HIGHBYTE\address.LOWBYTE,
[eeprom_data.HIGHBYTE]
  address = address + 1
  I2CIN EEPROM, EEPROM_READ, address.HIGHBYTE\address.LOWBYTE,
[eeprom_data.LOWBYTE]
  address = address + 1
  DEBUG DEC eeprom_data, TAB
  I2CIN EEPROM, EEPROM_READ, address.HIGHBYTE\address.LOWBYTE,
[eeprom_data.HIGHBYTE]
  address = address + 1
  I2CIN EEPROM, EEPROM_READ, address.HIGHBYTE\address.LOWBYTE,
[eeprom_data.LOWBYTE]
  address = address + 1
  DEBUG DEC eeprom_data, CR
LOOP WHILE address < EEPROM_SIZE
END
```

CODE 7

```
=====================================================================
'    Project 43
'    Accelerometer.bas
'    {$STAMP BS2p}
'    {$PBASIC 2.5}
=====================================================================
' -----[ Program Description ]----------------------------------------
' This program reads data from an accelerometer via an ADC, and stores the
' data in external EEPROM until the EEPROM is full.
' --------------------------------------------------------------------
' -----[ Pin assignments ]--------------------------------------------
G_SWITCH      PIN     1          ' Acceleration switch
ADC_SCLK      PIN     0          ' ADC serial clock pin
ADC_CS        PIN     1          ' ADC chip select pin
ADC_DATA      PIN     2          ' ADC serial data pin
EEPROM        PIN     0          ' EEPROM data pin is 0; EEPROM clock pin is 1
' -----[ Constant definitions ]---------------------------------------
EEPROM_WRITE CON      %10100000  ' Control code to perform write operation to
                                   I2C EEPROM
EEPROM_SIZE  CON      32768      ' Size in bytes of EEPROM
' -----[ Variable definitions ]---------------------------------------
adc_value     VAR     Word       ' Value read from ADC
address       VAR     Word       ' Address in EEPROM to write data
eeprom_data   VAR     Byte       ' Data byte to be written to EEPROM
' -----[ Initialisation ]---------------------------------------------
INPUT G_SWITCH                    ' Sets g-switch pin to input
HIGH ADC_CS                       ' Switches off ADC
address = 0
' -----[ Main program ]-----------------------------------------------
Wait_for_liftoff:
  IF G_SWITCH=0 THEN wait_for_liftoff    ' Waiting for lift off to be detected
DO
  GOSUB Read_ADC
```

```
    eeprom_data = adc_value.HIGHBYTE

    GOSUB Write_EEPROM

    eeprom_data = adc_value.LOWBYTE

    GOSUB Write_EEPROM

LOOP WHILE address < EEPROM_SIZE

END

' -----[ Subroutines ]---------------------------------------------------------

Read_ADC:
    LOW ADC_CS
    SHIFTIN ADC_DATA,ADC_SCLK,MSBPOST,[adc_value\12]
    HIGH ADC_CS

RETURN

Write_EEPROM:
    I2COUT EEPROM, EEPROM_WRITE, address.HIGHBYTE\address.LOWBYTE, [eeprom_data]
    address = address + 1

RETURN
```

CODE 8

```
===============================================================
'    Project 44
'    Vibration.bas
'    {$STAMP BS2p}
'    {$PBASIC 2.5}

===============================================================
' -----[ Program Description ]----------------------------------
' This program reads vibration data from a two-axis accelerometer via an
' ADC, and then stores the data in external EEPROM until the EEPROM is
' full.
'
' --------------------------------------------------------------

' -----[ Pin assignments ]--------------------------------------
G_SWITCH     PIN     1          ' Acceleration switch
ADC_SCLK     PIN     0          ' ADC serial clock pin
ADC_CS       PIN     1          ' ADC chip select pin
ADC_DATAOUT  PIN     2          ' Stamp -> ADC serial data pin
ADC_DATAIN   PIN     3          ' ADC -> Stamp serial data pin
EEPROM       PIN     0          ' EEPROM data pin is 0; EEPROM clock pin is 1

' -----[ Constant definitions ]---------------------------------
VIBRATION_X CON     142         ' X-axis on ADC channel 0
VIBRATION_Y CON     206         ' Y-axis on ADC channel 1
EEPROM_WRITECON     %10100000   'Control code to perform write operation to
                                 I2C EEPROM

EEPROM_SIZE  CON    32768       ' Size in bytes of EEPROM

' -----[ Variable definitions ]---------------------------------
adc_value    VAR     Word       ' Value read from ADC
channel      VAR     Byte       ' Analog channel to read from

address      VAR     Word       ' Address in EEPROM to write data
eeprom_data VAR      Byte       ' Data byte to be written to EEPROM

' -----[ Initialisation ]---------------------------------------
INPUT G_SWITCH                  ' Sets g-switch pin to input
HIGH ADC_CS                     ' Switches off ADC

address = 0

' -----[ Main program ]-----------------------------------------
Wait_for_liftoff:
   IF G_SWITCH=0 THEN wait_for_liftoff    ' Waiting for lift off to be detected
```

```
          DO
            channel = VIBRATION_X
            GOSUB Read_ADC
            eeprom_data = adc_value.HIGHBYTE
            GOSUB Write_EEPROM
            eeprom_data = adc_value.LOWBYTE
            GOSUB Write_EEPROM
            channel = VIBRATION_Y
            GOSUB Read_ADC
            eeprom_data = adc_value.HIGHBYTE
            GOSUB Write_EEPROM
            eeprom_data = adc_value.LOWBYTE
            GOSUB Write_EEPROM
          LOOP WHILE address < EEPROM_SIZE

          END

' -----[ Subroutines ]-----------------------------------------------------

Read_ADC:

    LOW ADC_CS
    SHIFTOUT ADC_DATAOUT, ADC_SCLK, MSBFIRST, [channel]
    SHIFTIN ADC_DATAIN, ADC_SCLK, MSBPOST, [adc_value\12]
    HIGH ADC_CS

RETURN


Write_EEPROM:
    I2COUT EEPROM, EEPROM_WRITE, address.HIGHBYTE\address.LOWBYTE, [eeprom_data]
    address = address + 1

RETURN
```

CODE 9

```
================================================================
'    Project 45
'    RollRate.bas
'    {$STAMP BS2p}
'    {$PBASIC 2.5}

================================================================
' -----[ Pin assignments ]----------------------------------------
G_SWITCH      PIN    1              ' Acceleration switch
LIGHT_IN      PIN    2              ' Light -> frequency input
EEPROM        PIN    0              ' EEPROM data pin is 0; EEPROM clock pin is 1
' -----[ Constant definitions ]-----------------------------------
EEPROM_WRITE CON    %10100000       ' Control code to perform write operation to
                                      I2C EEPROM
EEPROM_SIZE  CON    32768           ' Size in bytes of EEPROM
' -----[ Variable definitions ]-----------------------------------
light_level  VAR    Word
address      VAR    Word            ' Address in EEPROM to write data
eeprom_data  VAR    Byte            ' Data byte to be written to EEPROM
' -----[ Initialisation ]-----------------------------------------
INPUT G_SWITCH                       ' Sets g-switch pin to input
address = 0
' -----[ Main program ]-------------------------------------------
Wait_for_liftoff:
  IF G_SWITCH=0 THEN wait_for_liftoff    ' Waiting for lift off to be detected
DO
  COUNT LIGHT_IN, 10, light_level         ' Count pulses on LIGHT_IN for 10ms
                                          ' and store in light_level

  eeprom_data = light_level.HIGHBYTE
  GOSUB Write_EEPROM

  eeprom_data = light_level.LOWBYTE
  GOSUB Write_EEPROM
LOOP WHILE address < EEPROM_SIZE
END
' -----[ Subroutines ]--------------------------------------------
Write_EEPROM:
  I2COUT EEPROM, EEPROM_WRITE, address.HIGHBYTE\address.LOWBYTE, [eeprom_data]
  address = address + 1
RETURN
```

```
=========================================================================
'

'     Project 46
'     Thermocouple.bas
'     {$STAMP BS2p}
'     {$PBASIC 2.5}


=========================================================================
' -----[ Program Description ]--------------------------------------------
' This program reads Seebeck voltage and cold junction temperature from the
' DS2760 One-Wire thermocouple interface. This data is stored for later
' processing in external EEPROM.

  -------------------------------------------------------------------------
' -----[ Pin assignments ]------------------------------------------------
G_SWITCH        PIN     1       ' Acceleration switch
TC              PIN     2       ' One -Wire interface to thermocouple
EEPROM          PIN     0       ' EEPROM data pin is 0; EEPROM clock pin is 1

' -----[ Constant definitions ]-------------------------------------------
SkipNetAddr     CON     $CC     ' Control code to skip One-Wire net address
ReadReg         CON     $69     ' Control code to read register
EEPROM_WRITE    CON     %10100000 ' Control code to perform write operation to
                                ' I2C EEPROM

EEPROM_SIZE     CON     32768   ' Size in bytes of EEPROM

' -----[ Variable definitions ]-------------------------------------------
SeebeckV        VAR     Word    ' Seebeck voltage read from thermocouple
sign            VAR     Bit     ' Sign of Seebeck voltage
CJ_temp         VAR     Word    ' Cold junction temperature
address         VAR     Word    ' Address in EEPROM to write data
eeprom_data     VAR     Byte    ' Data byte to be written to EEPROM

' -----[ Initialisation ]-------------------------------------------------
INPUT G_SWITCH                  ' Sets g-switch pin to input
address = 0
' -----[ Main program ]---------------------------------------------------
wait_for_liftoff:
  IF G_SWITCH=0 THEN wait_for_liftoff     ' waiting for lift off to be detected
DO

  GOSUB Read_TC_Voltage
```

```
    eeprom_data = SeebeckV.HIGHBYTE
    GOSUB Write_EEPROM
    eeprom_data = SeebeckV.LOWBYTE
    GOSUB Write_EEPROM
    GOSUB Read_CJ_Temp
    eeprom_data = CJ_temp.HIGHBYTE
    GOSUB Write_EEPROM
    eeprom_data = CJ_temp.LOWBYTE
    GOSUB Write_EEPROM
LOOP WHILE address < EEPROM_SIZE
END


' -----[ Subroutines ]-------------------------------------------------

Read_TC_Voltage:
    OWOUT TC, %0001, [SkipNetAddr, ReadReg, $0E]
    OWIN  TC, %0010, [SeebeckV.HIGHBYTE, SeebeckV.LOWBYTE]

    sign = SeebeckV.BIT15                    ' save sign bit
    SeebeckV = SeebeckV >> 3                 ' correct alignment

    IF sign THEN
      SeebeckV = SeebeckV | $F000            ' pad 2's-compliment bits
    ENDIF

    SeebeckV = ABS SeebeckV */ 4000          ' x 15.625 uV
RETURN

Read_CJ_Temp:
    OWOUT TC, %0001, [SkipNetAddr, ReadReg, $18]
    OWIN  TC, %0010, [CJ_temp.BYTE1, CJ_temp.BYTE0]
    IF (CJ_temp.BIT15) THEN                  ' check sign
      CJ_temp = 0                            ' disallow negative
    ELSE
      CJ_temp = CJ_temp.HIGHBYTE             ' >> 5 x 0.125 (>> 3)
    ENDIF
RETURN

Write_EEPROM:
    I2COUT EEPROM, EEPROM_WRITE, address.HIGHBYTE\address.LOWBYTE, [eeprom_data]
    address = address + 1
RETURN
```

```
' ==============================================================
'    Project 47
'    DataDownloadTC.bas
'    {$STAMP BS2p, KTablePos.BPE, JTablePos.BPE}
'    {$PBASIC 2.5}
' ==============================================================
' -----[ Program Description ]----------------------------------
' This program...
' --------------------------------------------------------------
' -----[ Pin assignments ]--------------------------------------

EEPROM          PIN    0            ' EEPROM data pin is 0; EEPROM clock pin is 1
' -----[ Constant definitions ]---------------------------------
SkipNetAddr     CON    $CC          ' Control code to skip One-Wire net address
ReadReg         CON    $69          ' Control code to read register
EEPROM_READ     CON    %10100001    ' Control code to perform read operation to
                                    ' I2C EEPROM
EEPROM_SIZE     CON    32768        ' Size in bytes of EEPROM

' -----[ Variable definitions ]---------------------------------
type            VAR    Byte         ' Type of thermocouple - either K or J
address         VAR    Word         ' Address in EEPROM to write data
eeprom_data     VAR    Byte         ' Data byte to be written to EEPROM

SeebeckV        VAR    Word         ' Seebeck voltage read from thermocouple
CJ_temp         VAR    Word         ' Cold junction temperature
CJ_compV        VAR    Word         ' Compensation voltage
Temp            VAR    Word         ' Thermocouple temperature in Celsius
error           VAR    Bit          ' Error bit is set if temperature is too high

_hi_pntr        VAR    Word         ' Table pointers
_lo_pntr        VAR    Word         '
_mid_pntr       VAR    Word         '
test_val        VAR    Word         ' Value to test compensation voltage against

' -----[ Initialisation ]---------------------------------------
address = 0
_hi_pntr = 1023

Select_TC_Type:
  DEBUG CLS,
```

```
            "Select TC Type", CR,
            CR,
            "K - Chromel/Alumel", CR,
            "J - Iron/Constantan", CR,
            CR,
            ">>> "
    DEBUGIN type
    IF (type <> "K") OR (type <> "J") THEN Select_TC_Type
    DEBUG CLS
    IF (type = "K") THEN
      type = 1
    ELSE
      type = 2
    ENDIF
    STORE type                                      ' point READ to table
' -----[ Main program ]-------------------------------------------------
DO
  GOSUB Read_EEPROM
  SeebeckV.HIGHBYTE = eeprom_data
  GOSUB Read_EEPROM
  SeebeckV.LOWBYTE = eeprom_data
  GOSUB Read_EEPROM
  CJ_temp.HIGHBYTE = eeprom_data
  GOSUB Read_EEPROM
  CJ_temp.LOWBYTE = eeprom_data
  READ (CJ_temp * 2), Word CJ_compV            ' get CJ compensation voltage
  CJ_compV = CJ_compV + SeebeckV
  GOSUB TC_Lookup
  IF (error = 0) THEN
    DEBUG DEC Temp, CR
  ELSE
    DEBUG "Out of range", CR
  ENDIF
LOOP WHILE address < EEPROM_SIZE
END
' -----[ Subroutines ]--------------------------------------------------
Read_EEPROM:
  I2CIN EEPROM, EEPROM_READ, address.HIGHBYTE\address.LOWBYTE, [eeprom_data]
  address = address + 1
RETURN

TC_Lookup:
  _lo_pntr = 0
  READ (_hi_pntr * 2), Word test_val                ' check max temp
```

```
      IF (CJ_compV > test_val) THEN
        error = 1                                   ' out of range
      ELSE
        DO
          _mid_pntr = (_lo_pntr + _hi_pntr) / 2       ' midpoint of search
span
          READ (_mid_pntr * 2), Word test_val       ' read value from midpoint
          IF (CJ_compV = test_val) THEN
            EXIT                                    ' found it!
          ELSEIF (CJ_compV < test_val) THEN
            _hi_pntr = _mid_pntr                       ' search lower half
          ELSE
            _lo_pntr = _mid_pntr                       ' search upper half
          ENDIF
          IF ((_hi_pntr - _lo_pntr) < 2) THEN          ' span at minimum
            _mid_pntr = _lo_pntr
            EXIT
          ENDIF
        LOOP
        temp = _mid_pntr
      ENDIF
RETURN
```

```
=================================================================
'    Project 47
'    DataAcquisitionSystem.bas
'    {$STAMP BS2p}
'    {$PBASIC 2.5}

=================================================================
' -----[ Pin assignments ]----------------------------------------
G_SWITCH        PIN    1              ' Acceleration switch
ADC_SCLK        PIN    0              ' ADC serial clock pin
ADC_CS          PIN    1              ' ADC chip select pin
ADC_DATAOUT     PIN    2              ' Stamp -> ADC serial data pin
ADC_DATAIN      PIN    3              ' ADC -> Stamp serial data pin
TH_DATA         PIN    2              ' Sensor data pin
TH_CLOCK        PIN    3              ' Sensor clock pin
EEPROM          PIN    0              ' EEPROM data pin is 0; EEPROM clock pin is 1

' -----[ Constant definitions ]-----------------------------------
' ADC channels
BARO            CON    142            ' Channel 0 : Air pressure sensor
ACCEL           CON    206            ' Channel 1 : Acceleromter
VIB_X           CON                   ' Channel 2 : Vibration X-axis
VIB_Y           CON                   ' Channel 3 : Vibration Y-axis
ROLL            CON                   ' Channel 4 : Gyroscope
MAG_FIELD       CON                   ' Channel 5 : Magnetic field sensor
TEMP            CON    %00011         ' Control codes to read temperature and
HUMID           CON    %00101         ' humidity from SHT11 sensor
STATUS_WRITE    CON    %00110         ' Control codes to read and write SHT11
STATUS_READ     CON    %00111         ' status register

EEPROM_WRITE    CON    %10100000      ' Control code to perform write operation to
                                      I2C EEPROM
EEPROM_SIZE     CON    32768          ' Size in bytes of EEPROM

' -----[ Variable definitions ]-----------------------------------
adc_value   VAR    Word        ' Value read from ADC
channel     VAR    Byte        ' Analog channel to read from
address     VAR    Word        ' Address in EEPROM to write data
eeprom_data VAR    Byte        ' Data byte to be written to EEPROM

' -----[ Initialisation ]----------------------------------------
INPUT G_SWITCH                  ' Sets g-switch pin to input
HIGH ADC_CS                     ' Switches off ADC
SHIFTOUT TH_DATA, TH_CLOCK, LSBFIRST, [$FFF\9]
GOSUB TH_Start_Sequence         ' Resets connection to temp/humidity sensor
address = 0
```

```
' -----[ Main program ]-------------------------------------------------
Wait_for_liftoff:
  IF G_SWITCH=0 THEN wait_for_liftoff      ' Waiting for lift off to be detected
DO
  channel = BARO
  GOSUB Read_ADC
  eeprom_data = adc_value.HIGHBYTE

  GOSUB Write_EEPROM
  eeprom_data = adc_value.LOWBYTE

  GOSUB Write_EEPROM
  channel = ACCEL

  GOSUB Read_ADC
  eeprom_data = adc_value.HIGHBYTE

  GOSUB Write_EEPROM
  eeprom_data = adc_value.LOWBYTE

  GOSUB Write_EEPROM
  channel = VIB_X

  GOSUB Read_ADC
  eeprom_data = adc_value.HIGHBYTE

  GOSUB Write_EEPROM
  eeprom_data = adc_value.LOWBYTE

  GOSUB Write_EEPROM
  channel = VIB_Y

  GOSUB Read_ADC
  eeprom_data = adc_value.HIGHBYTE

  GOSUB Write_EEPROM
  eeprom_data = adc_value.LOWBYTE

  GOSUB Write_EEPROM
LOOP WHILE address < EEPROM_SIZE
END
' -----[ Subroutines ]-------------------------------------------------
Read_ADC:
    LOW ADC_CS
    SHIFTOUT ADC_DATAOUT, ADC_SCLK, MSBFIRST, [channel]
    SHIFTIN ADC_DATAIN, ADC_SCLK, MSBPOST, [adc_value\12]
    HIGH ADC_CS
RETURN
Write_EEPROM:
  I2COUT EEPROM, EEPROM_WRITE, address.HIGHBYTE\address.LOWBYTE, [eeprom_data]
  address = address + 1
RETURN
```

CODE 13

```
=================================================================
'
'    DataDownload8.bas
'    {$STAMP BS2p}
'    {$PBASIC 2.5}
=================================================================
' -----[ Program Description ]------------------------------------
' This program...
' ----------------------------------------------------------------
' -----[ Pin assignments ]----------------------------------------
EEPROM          PIN     0               ' EEPROM data pin is 0; EEPROM clock pin is 1
' -----[ Constant definitions ]-----------------------------------
EEPROM_READ     CON     %10100001       ' Control code to perform read operation to
                                        '   I2C EEPROM
EEPROM_SIZE     CON     32768           ' Size in bytes of EEPROM
' -----[ Variable definitions ]-----------------------------------
address         VAR     Word            ' Address in EEPROM to write data
eeprom_data     VAR     Word            ' Data byte to be written to EEPROM

' -----[ Initialisation ]-----------------------------------------
address = 0

' -----[ Main program ]-------------------------------------------
DO
   I2CIN EEPROM, EEPROM_READ, address.HIGHBYTE\address.LOWBYTE,
[eeprom_data.HIGHBYTE]
   address = address + 1
   I2CIN EEPROM, EEPROM_READ, address.HIGHBYTE\address.LOWBYTE,
[eeprom_data.LOWBYTE]
   address = address + 1
   DEBUG DEC eeprom_data, TAB

   I2CIN EEPROM, EEPROM_READ, address.HIGHBYTE\address.LOWBYTE,
[eeprom_data.HIGHBYTE]
   address = address + 1
   I2CIN EEPROM, EEPROM_READ, address.HIGHBYTE\address.LOWBYTE,
[eeprom_data.LOWBYTE]
   address = address + 1
   DEBUG DEC eeprom_data, TAB

[eeprom_data.HIGHBYTE]
   address = address + 1
```

```
    I2CIN EEPROM, EEPROM_READ, address.HIGHBYTE\address.LOWBYTE,
[eeprom_data.LOWBYTE]
    address = address + 1
    DEBUG DEC eeprom_data, TAB
    I2CIN EEPROM, EEPROM_READ, address.HIGHBYTE\address.LOWBYTE,
[eeprom_data.HIGHBYTE]
    address = address + 1

    I2CIN EEPROM, EEPROM_READ, address.HIGHBYTE\address.LOWBYTE,
[eeprom_data.LOWBYTE]
    address = address + 1
    DEBUG DEC eeprom_data, TAB
    I2CIN EEPROM, EEPROM_READ, address.HIGHBYTE\address.LOWBYTE,
[eeprom_data.HIGHBYTE]
    address = address + 1

    I2CIN EEPROM, EEPROM_READ, address.HIGHBYTE\address.LOWBYTE,
[eeprom_data.LOWBYTE]
    address = address + 1
    DEBUG DEC eeprom_data, TAB
    I2CIN EEPROM, EEPROM_READ, address.HIGHBYTE\address.LOWBYTE,
[eeprom_data.HIGHBYTE]
    address = address + 1

    I2CIN EEPROM, EEPROM_READ, address.HIGHBYTE\address.LOWBYTE,
[eeprom_data.LOWBYTE]
    address = address + 1
    DEBUG DEC eeprom_data, TAB
    I2CIN EEPROM, EEPROM_READ, address.HIGHBYTE\address.LOWBYTE,
[eeprom_data.HIGHBYTE]
    address = address + 1

    I2CIN EEPROM, EEPROM_READ, address.HIGHBYTE\address.LOWBYTE,
[eeprom_data.LOWBYTE]
    address = address + 1
    DEBUG DEC eeprom_data, TAB
    I2CIN EEPROM, EEPROM_READ, address.HIGHBYTE\address.LOWBYTE,
[eeprom_data.HIGHBYTE]
    address = address + 1
    DEBUG DEC eeprom_data, CR
LOOP WHILE address < EEPROM_SIZE

END
```

Appendix 12-2. Lookup Tables (courtesy Parallax Inc)

"J" type thermocouple

Table 12-1

```
=================================================================
```
```
' File...... JTablePos.BPE
' Purpose... J-type (Iron/Constantan) thermocouple data (0C reference)
' Author.... Compiled by Parallax
' E-mail.... support@parallax.com
' Started...

' Updated... 19 JAN 2004
' {$STAMP BS2pe}
' {$PBASIC 2.5}
```
```
=================================================================
```

tC	+0	+1	+2	+3	+4
	+5	+6	+7	+8	+9
J0000 DATA	Word 00000,	Word 00050,	Word 00101,	Word 00151,	Word 00202,
	Word 00253,	Word 00303,	Word 00354,	Word 00405,	Word 00456
J0010 DATA	Word 00507,	Word 00558,	Word 00609,	Word 00660,	Word 00711,
	Word 00762,	Word 00814,	Word 00865,	Word 00916,	Word 00968
J0020 DATA	Word 01019,	Word 01070,	Word 01122,	Word 01174,	Word 01226,
	Word 01277,	Word 01328,	Word 01381,	Word 01433,	Word 01485
J0030 DATA	Word 01536,	Word 01588,	Word 01641,	Word 01693,	Word 01745,
	Word 01796,	Word 01848,	Word 01902,	Word 01954,	Word 02006
J0040 DATA	Word 02059,	Word 02111,	Word 02164,	Word 02216,	Word 02269,
	Word 02322,	Word 02374,	Word 02427,	Word 02480,	Word 02532
J0050 DATA	Word 02584,	Word 02637,	Word 02690,	Word 02744,	Word 02797,
	Word 02850,	Word 02903,	Word 02956,	Word 03008,	Word 03062
J0060 DATA	Word 03116,	Word 03169,	Word 03222,	Word 03274,	Word 03329,
	Word 03382,	Word 03435,	Word 03488,	Word 03543,	Word 03596
J0070 DATA	Word 03649,	Word 03703,	Word 03757,	Word 03810,	Word 03863,
	Word 03918,	Word 03971,	Word 04025,	Word 04078,	Word 04133
J0080 DATA	Word 04187,	Word 04240,	Word 04294,	Word 04347,	Word 04402,
	Word 04456,	Word 04509,	Word 04564,	Word 04618,	Word 04671
J0090 DATA	Word 04726,	Word 04780,	Word 04835,	Word 04889,	Word 04943,
	Word 04996,	Word 05052,	Word 05105,	Word 05160,	Word 05214
J0100 DATA	Word 05269,	Word 05323,	Word 05378,	Word 05432,	Word 05487,
	Word 05541,	Word 05594,	Word 05650,	Word 05705,	Word 05759
J0110 DATA	Word 05814,	Word 05868,	Word 05923,	Word 05977,	Word 06032,
	Word 06086,	Word 06141,	Word 06195,	Word 06251,	Word 06306
J0120 DATA	Word 06360,	Word 06415,	Word 06469,	Word 06525,	Word 06578,
	Word 06634,	Word 06689,	Word 06743,	Word 06799,	Word 06854
J0130 DATA	Word 06908,	Word 06964,	Word 07019,	Word 07073,	Word 07129,
	Word 07184,	Word 07238,	Word 07294,	Word 07349,	Word 07403

| tC | +0 | +1 | +2 | +3 | +4 |
	+5	+6	+7	+8	+9
J0140 DATA	Word 07459,	Word 07514,	Word 07569,	Word 07624,	Word 07679,
	Word 07734,	Word 07789,	Word 07844,	Word 07900,	Word 07955
J0150 DATA	Word 08009,	Word 08064,	Word 08120,	Word 08175,	Word 08230,
	Word 08285,	Word 08340,	Word 08396,	Word 08451,	Word 08506
J0160 DATA	Word 08561,	Word 08618,	Word 08673,	Word 08727,	Word 08782,
	Word 08839,	Word 08894,	Word 08948,	Word 09005,	Word 09060
J0170 DATA	Word 09115,	Word 09170,	Word 09226,	Word 09282,	Word 09336,
	Word 09391,	Word 09448,	Word 09503,	Word 09559,	Word 09614
J0180 DATA	Word 09669,	Word 09724,	Word 09779,	Word 09836,	Word 09891,
	Word 09947,	Word 10002,	Word 10057,	Word 10112,	Word 10168
J0190 DATA	Word 10224,	Word 10278,	Word 10335,	Word 10390,	Word 10445,
	Word 10500,	Word 10557,	Word 10612,	Word 10668,	Word 10723
J0200 DATA	Word 10778,	Word 10833,	Word 10890,	Word 10945,	Word 11000,
	Word 11056,	Word 11112,	Word 11166,	Word 11223,	Word 11278
J0210 DATA	Word 11333,	Word 11389,	Word 11445,	Word 11500,	Word 11556,
	Word 11612,	Word 11666,	Word 11723,	Word 11778,	Word 11833
J0220 DATA	Word 11889,	Word 11945,	Word 12000,	Word 12056,	Word 12111,
	Word 12166,	Word 12221,	Word 12278,	Word 12333,	Word 12389
J0230 DATA	Word 12445,	Word 12500,	Word 12556,	Word 12611,	Word 12666,
	Word 12721,	Word 12778,	Word 12833,	Word 12889,	Word 12944
J0240 DATA	Word 13000,	Word 13056,	Word 13111,	Word 13166,	Word 13221,
	Word 13278,	Word 13333,	Word 13389,	Word 13444,	Word 13500
J0250 DATA	Word 13554,	Word 13611,	Word 13666,	Word 13721,	Word 13777,
	Word 13833,	Word 13887,	Word 13944,	Word 13999,	Word 14054
J0260 DATA	Word 14109,	Word 14166,	Word 14221,	Word 14277,	Word 14332,
	Word 14387,	Word 14442,	Word 14499,	Word 14554,	Word 14608
J0270 DATA	Word 14665,	Word 14720,	Word 14775,	Word 14830,	Word 14887,
	Word 14942,	Word 14998,	Word 15053,	Word 15108,	Word 15163
J0280 DATA	Word 15219,	Word 15275,	Word 15330,	Word 15386,	Word 15441,
	Word 15496,	Word 15551,	Word 15607,	Word 15663,	Word 15718
J0290 DATA	Word 15772,	Word 15829,	Word 15884,	Word 15939,	Word 15995,
	Word 16050,	Word 16106,	Word 16161,	Word 16216,	Word 16272
J0300 DATA	Word 16327,	Word 16382,	Word 16437,	Word 16493,	Word 16548,
	Word 16603,	Word 16658,	Word 16714,	Word 16769,	Word 16824
J0310 DATA	Word 16881,	Word 16935,	Word 16990,	Word 17045,	Word 17102,
	Word 17157,	Word 17211,	Word 17268,	Word 17323,	Word 17378
J0320 DATA	Word 17434,	Word 17489,	Word 17544,	Word 17599,	Word 17655,
	Word 17710,	Word 17765,	Word 17820,	Word 17876,	Word 17931
J0330 DATA	Word 17986,	Word 18041,	Word 18097,	Word 18152,	Word 18207,
	Word 18262,	Word 18318,	Word 18373,	Word 18428,	Word 18483
J0340 DATA	Word 18538,	Word 18594,	Word 18649,	Word 18704,	Word 18759,
	Word 18814,	Word 18870,	Word 18925,	Word 18980,	Word 19035
J0350 DATA	Word 19089,	Word 19146,	Word 19201,	Word 19256,	Word 19310,
	Word 19365,	Word 19422,	Word 19477,	Word 19532,	Word 19586
J0360 DATA	Word 19641,	Word 19696,	Word 19753,	Word 19807,	Word 19862,
	Word 19917,	Word 19972,	Word 20027,	Word 20083,	Word 20138

tC	+0 / +5	+1 / +6	+2 / +7	+3 / +8	+4 / +9
J0370 DATA	Word 20193,	Word 20248,	Word 20304,	Word 20359,	Word 20414,
	Word 20469,	Word 20525,	Word 20580,	Word 20635,	Word 20690
J0380 DATA	Word 20745,	Word 20800,	Word 20855,	Word 20911,	Word 20966,
	Word 21021,	Word 21076,	Word 21131,	Word 21186,	Word 21240
J0390 DATA	Word 21297,	Word 21352,	Word 21407,	Word 21461,	Word 21516,
	Word 21571,	Word 21626,	Word 21682,	Word 21737	Word 21792
J0400 DATA	Word 21847,	Word 21902,	Word 21958,	Word 22013,	Word 22068,
	Word 22123,	Word 22179,	Word 22234,	Word 22289,	Word 22344
J0410 DATA	Word 22400,	Word 22455,	Word 22510,	Word 22565,	Word 22620,
	Word 22676,	Word 22731,	Word 22786,	Word 22841,	Word 22896
J0420 DATA	Word 22952,	Word 23007,	Word 23062,	Word 23117,	Word 23172,
	Word 23228,	Word 23283,	Word 23338,	Word 23393,	Word 23449
J0430 DATA	Word 23504,	Word 23559,	Word 23614,	Word 23670,	Word 23725,
	Word 23780,	Word 23835,	Word 23891,	Word 23946,	Word 24001
J0440 DATA	Word 24057,	Word 24112,	Word 24167,	Word 24222,	Word 24278,
	Word 24333,	Word 24388,	Word 24443,	Word 24498,	Word 24554
J0450 DATA	Word 24609,	Word 24664,	Word 24721,	Word 24775,	Word 24832,
	Word 24887,	Word 24943,	Word 24998,	Word 25053,	Word 25109
J0460 DATA	Word 25164,	Word 25219,	Word 25275,	Word 25330,	Word 25385,
	Word 25442,	Word 25496,	Word 25553,	Word 25608,	Word 25664
J0470 DATA	Word 25719,	Word 25775,	Word 25830,	Word 25885,	Word 25942,
	Word 25998,	Word 26053,	Word 26109,	Word 26164,	Word 26219
J0480 DATA	Word 26275,	Word 26332,	Word 26387,	Word 26443,	Word 26498,
	Word 26554,	Word 26609,	Word 26666,	Word 26722,	Word 26778
J0490 DATA	Word 26833,	Word 26888,	Word 26945,	Word 27001,	Word 27057,
	Word 27112,	Word 27169,	Word 27225,	Word 27280,	Word 27336
J0500 DATA	Word 27393,	Word 27449,	Word 27504,	Word 27561,	Word 27617,
	Word 27673,	Word 27728,	Word 27785,	Word 27841,	Word 27897
J0510 DATA	Word 27952,	Word 28010,	Word 28065,	Word 28121,	Word 28178,
	Word 28234,	Word 28291,	Word 28347,	Word 28403,	Word 28460
J0520 DATA	Word 28516,	Word 28571,	Word 28629,	Word 28685,	Word 28740,
	Word 28798,	Word 28853,	Word 28911,	Word 28967,	Word 29024
J0530 DATA	Word 29080,	Word 29137,	Word 29193,	Word 29250,	Word 29307,
	Word 29362,	Word 29420,	Word 29477,	Word 29533,	Word 29589
J0540 DATA	Word 29647,	Word 29704,	Word 29760,	Word 29818,	Word 29874,
	Word 29931,	Word 29987,	Word 30045,	Word 30102,	Word 30158
J0550 DATA	Word 30216,	Word 30272,	Word 30330,	Word 30387,	Word 30443,
	Word 30501,	Word 30559,	Word 30615,	Word 30673,	Word 30730
J0560 DATA	Word 30788,	Word 30845,	Word 30902,	Word 30960,	Word 31016,
	Word 31074,	Word 31132,	Word 31189,	Word 31246,	Word 31304
J0570 DATA	Word 31362,	Word 31419,	Word 31477,	Word 31535,	Word 31592
J0580 DATA	Word 31939,	Word 31996,	Word 32054,	Word 32112,	Word 32170,
	Word 32228,	Word 32286,	Word 32344,	Word 32402,	Word 32460
J0590 DATA	Word 32518,	Word 32576,	Word 32636,	Word 32694,	Word 32752,
	Word 32810,	Word 32868,	Word 32926,	Word 32984,	Word 33044
J0600 DATA	Word 33102,	Word 33161,	Word 33219,	Word 33277,	Word 33337,
	Word 33395,	Word 33454,	Word 33512,	Word 33570,	Word 33630

| tC | +0 | +1 | +2 | +3 | +4 |
	+5	+6	+7	+8	+9
J0610 DATA	Word 33689,	Word 33747,	Word 33807,	Word 33865,	Word 33924,
	Word 33984,	Word 34042,	Word 34102,	Word 34161,	Word 34219
J0620 DATA	Word 34279,	Word 34338,	Word 34396,	Word 34457,	Word 34515,
	Word 34575,	Word 34634,	Word 34694,	Word 34753,	Word 34813
J0630 DATA	Word 34872,	Word 34932,	Word 34991,	Word 35051,	Word 35111,
	Word 35170,	Word 35230,	Word 35289,	Word 35350,	Word 35410
J0640 DATA	Word 35469,	Word 35530,	Word 35590,	Word 35649,	Word 35710,
	Word 35770,	Word 35829,	Word 35890,	Word 35950,	Word 36009
J0650 DATA	Word 36070,	Word 36131,	Word 36191,	Word 36252,	Word 36311,
	Word 36372,	Word 36432,	Word 36493,	Word 36554,	Word 36615
J0660 DATA	Word 36675,	Word 36736,	Word 36797,	Word 36858,	Word 36917,
	Word 36978,	Word 37039,	Word 37100,	Word 37161,	Word 37222
J0670 DATA	Word 37283,	Word 37344,	Word 37405,	Word 37466,	Word 37527,
	Word 37590,	Word 37651,	Word 37712,	Word 37773,	Word 37835
J0680 DATA	Word 37896,	Word 37957,	Word 38018,	Word 38081,	Word 38142,
	Word 38204,	Word 38265,	Word 38326,	Word 38389,	Word 38450
J0690 DATA	Word 38512,	Word 38573,	Word 38636,	Word 38698,	Word 38759,
	Word 38822,	Word 38884,	Word 38945,	Word 39008,	Word 39070
J0700 DATA	Word 39132,	Word 39194,	Word 39256,	Word 39317,	Word 39381,
	Word 39442,	Word 39505,	Word 39567,	Word 39630,	Word 39692
J0710 DATA	Word 39755,	Word 39817,	Word 39880,	Word 39942,	Word 40005,
	Word 40067,	Word 40131,	Word 40192,	Word 40256,	Word 40319
J0720 DATA	Word 40381,	Word 40445,	Word 40508,	Word 40570,	Word 40633,
	Word 40695,	Word 40759,	Word 40822,	Word 40886,	Word 40948
J0730 DATA	Word 41012,	Word 41075,	Word 41137,	Word 41201,	Word 41265,
	Word 41328,	Word 41390,	Word 41454,	Word 41518,	Word 41581
J0740 DATA	Word 41645,	Word 41707,	Word 41771,	Word 41835,	Word 41899,
	Word 41962,	Word 42026,	Word 42090,	Word 42152,	Word 42216
J0750 DATA	Word 42280,	Word 42344,	Word 42408,	Word 42472,	Word 42536,
	Word 42599,	Word 42663,	Word 42727,	Word 42791,	Word 42855
J0760 DATA	Word 42919,	Word 42983,	Word 43047,	Word 43111,	Word 43175,
	Word 43239,	Word 43303,	Word 43367,	Word 43430,	Word 43494
J0770 DATA	Word 43558,	Word 43624,	Word 43688,	Word 43752,	Word 43817,
	Word 43881,	Word 43945,	Word 44009,	Word 44073,	Word 44139
J0780 DATA	Word 44203,	Word 44267,	Word 44332,	Word 44396,	Word 44460,
	Word 44524,	Word 44590,	Word 44655,	Word 44719,	Word 44783
J0790 DATA	Word 44847,	Word 44913,	Word 44977,	Word 45042,	Word 45106,
	Word 45170,	Word 45236,	Word 45301,	Word 45365,	Word 45429
J0800 DATA	Word 45494,	Word 45558,	Word 45624,	Word 45688,	Word 45753,
	Word 45817,	Word 45881,	Word 45947,	Word 46011,	Word 46076
J0810 DATA	Word 46140,	Word 46204,	Word 46270,	Word 46334,	Word 46399,
	Word 46463,	Word 46527,	Word 46593,	Word 46657,	Word 46722
J0820 DATA	Word 46786,	Word 46850,	Word 46914,	Word 46980,	Word 47044,
	Word 47109,	Word 47173,	Word 47237,	Word 47301,	Word 47367
J0830 DATA	Word 47431,	Word 47494,	Word 47560,	Word 47624,	Word 47688,
	Word 47753,	Word 47817,	Word 47881,	Word 47945,	Word 48009

Lookup Tables

tC	+0 +5	+1 +6	+2 +7	+3 +8	+4 +9
J0840 DATA	Word 48073, Word 48395,	Word 48137, Word 48459,	Word 48201, Word 48523,	Word 48267, Word 48587,	Word 48331, Word 48651
J0850 DATA	Word 48715, Word 49033,	Word 48779, Word 49097,	Word 48843, Word 49161,	Word 48907, Word 49225,	Word 48971, Word 49289
J0860 DATA	Word 49353, Word 49672,	Word 49417, Word 49734,	Word 49481, Word 49798,	Word 49544, Word 49862,	Word 49608, Word 49926
J0870 DATA	Word 49989, Word 50306,	Word 50051, Word 50368,	Word 50115, Word 50432,	Word 50179, Word 50495,	Word 50243, Word 50559
J0880 DATA	Word 50621, Word 50936,	Word 50685, Word 51000,	Word 50747, Word 51063,	Word 50810, Word 51125,	Word 50874, Word 51188
J0890 DATA	Word 51250, Word 51564,	Word 51314, Word 51627,	Word 51377, Word 51689,	Word 51439, Word 51752,	Word 51502, Word 51814
J0900 DATA	Word 51877, Word 52189,	Word 51939, Word 52250,	Word 52002, Word 52314,	Word 52064, Word 52375,	Word 52127, Word 52438
J0910 DATA	Word 52500, Word 52810,	Word 52561, Word 52871,	Word 52624, Word 52934,	Word 52686, Word 52996,	Word 52747, Word 53057
J0920 DATA	Word 53118, Word 53426,	Word 53181, Word 53489,	Word 53243, Word 53550,	Word 53304, Word 53612,	Word 53365, Word 53673
J0930 DATA	Word 53734, Word 54041,	Word 53795, Word 54102,	Word 53856, Word 54164,	Word 53919, Word 54225,	Word 53980, Word 54286
J0940 DATA	Word 54347, Word 54652,	Word 54408, Word 54713,	Word 54469, Word 54773,	Word 54530, Word 54834,	Word 54591, Word 54895
J0950 DATA	Word 54956, Word 55259,	Word 55015, Word 55319,	Word 55076, Word 55380,	Word 55137, Word 55439,	Word 55198, Word 55500
J0960 DATA	Word 55560, Word 55862,	Word 55621, Word 55923,	Word 55682, Word 55983,	Word 55742, Word 56042,	Word 55803, Word 56103
J0970 DATA	Word 56164, Word 56463,	Word 56224, Word 56524,	Word 56283, Word 56584,	Word 56344, Word 56643,	Word 56404, Word 56703
J0980 DATA	Word 56763, Word 57062,	Word 56823, Word 57121,	Word 56883, Word 57181,	Word 56942, Word 57240,	Word 57002, Word 57300
J0990 DATA	Word 57359, Word 57657,	Word 57419, Word 57716,	Word 57478, Word 57776,	Word 57538, Word 57835,	Word 57597, Word 57893
J1000 DATA	Word 57953, Word 58249,	Word 58013, Word 58309,	Word 58072, Word 58368,	Word 58131, Word 58427,	Word 58190, Word 58486
J1010 DATA	Word 58545, Word 58840,	Word 58604, Word 58899,	Word 58663, Word 58957,	Word 58722, Word 59016,	Word 58781, Word 59075
J1020 DATA	Word 59134,	Word 59193,	Word 59252,	Word 59310	

Table 12-2

```
============================================================
```

```
"   File...... KTablePos.BPE
'   Purpose... K-type (Chromel/Alumel) thermocouple data (0C reference)
'   Author.... Compiled by Parallax
'   E-mail.... support@parallax.com
'   Started...
'   Updated... 19 JAN 2004
'
'   {$STAMP BS2pe}
'   {$PBASIC 2.5}
'
```

```
'============================================================
```

'tC	+0	+1	+2	+3	+4
'	+5	+6	+7	+8	+9
K0000 DATA	Word 00000,	Word 00039,	Word 00079,	Word 00119,	Word 00158,
	Word 00198,	Word 00238,	Word 00277,	Word 00317,	Word 00357
K0010 DATA	Word 00397,	Word 00437,	Word 00477,	Word 00517,	Word 00557,
	Word 00597,	Word 00637,	Word 00677,	Word 00718,	Word 00758
K0020 DATA	Word 00798,	Word 00838,	Word 00879,	Word 00919,	Word 00960,
	Word 01000,	Word 01040,	Word 01080,	Word 01122,	Word 01163
K0030 DATA	Word 01203,	Word 01244,	Word 01284,	Word 01326,	Word 01366,
	Word 01407,	Word 01448,	Word 01489,	Word 01530,	Word 01570
K0040 DATA	Word 01612,	Word 01653,	Word 01694,	Word 01735,	Word 01776,
	Word 01816,	Word 01858,	Word 01899,	Word 01941,	Word 01982
K0050 DATA	Word 02023,	Word 02064,	Word 02105,	Word 02146,	Word 02188,
	Word 02230,	Word 02270,	Word 02311,	Word 02354,	Word 02395
K0060 DATA	Word 02436,	Word 02478,	Word 02519,	Word 02560,	Word 02601,
	Word 02644,	Word 02685,	Word 02726,	Word 02767,	Word 02810
K0070 DATA	Word 02850,	Word 02892,	Word 02934,	Word 02976,	Word 03016,
	Word 03059,	Word 03100,	Word 03141,	Word 03184,	Word 03225
K0080 DATA	Word 03266,	Word 03307,	Word 03350,	Word 03391,	Word 03432,
	Word 03474,	Word 03516,	Word 03557,	Word 03599,	Word 03640
K0090 DATA	Word 03681,	Word 03722,	Word 03765,	Word 03806,	Word 03847,
	Word 03888,	Word 03931,	Word 03972,	Word 04012,	Word 04054
K0100 DATA	Word 04096,	Word 04137,	Word 04179,	Word 04219,	Word 04261,
	Word 04303,	Word 04344,	Word 04384,	Word 04426,	Word 04468
K0110 DATA	Word 04509,	Word 04549,	Word 04591,	Word 04633,	Word 04674,
	Word 04714,	Word 04756,	Word 04796,	Word 04838,	Word 04878
K0120 DATA	Word 04919,	Word 04961,	Word 05001,	Word 05043,	Word 05083,
	Word 05123,	Word 05165,	Word 05206,	Word 05246,	Word 05288
K0130 DATA	Word 05328,	Word 05368,	Word 05410,	Word 05450,	Word 05490,
	Word 05532,	Word 05572,	Word 05613,	Word 05652,	Word 05693
K0140 DATA	Word 05735,	Word 05775,	Word 05815,	Word 05865,	Word 05895,
	Word 05937,	Word 05977,	Word 06017,	Word 06057,	Word 06097

Lookup Tables

| 'tC | +0 | +1 | +2 | +3 | +4 |
'	+5	+6	+7	+8	+9
K0150 DATA	Word 06137,	Word 06179,	Word 06219,	Word 06259,	Word 06299,
	Word 06339,	Word 06379,	Word 06419,	Word 06459,	Word 06500
K0160 DATA	Word 06540,	Word 06580,	Word 06620,	Word 06660,	Word 06700,
	Word 06740,	Word 06780,	Word 06820,	Word 06860,	Word 06900
K0170 DATA	Word 06940,	Word 06980,	Word 07020,	Word 07059,	Word 07099,
	Word 07139,	Word 07179,	Word 07219,	Word 07259,	Word 07299
K0180 DATA	Word 07339,	Word 07379,	Word 07420,	Word 07459,	Word 07500,
	Word 07540,	Word 07578,	Word 07618,	Word 07658,	Word 07698
K0190 DATA	Word 07738,	Word 07778,	Word 07819,	Word 07859,	Word 07899,
	Word 07939,	Word 07979,	Word 08019,	Word 08058,	Word 08099
K0200 DATA	Word 08137,	Word 08178,	Word 08217,	Word 08257,	Word 08298,
	Word 08337,	Word 08378,	Word 08417,	Word 08458,	Word 08499
K0210 DATA	Word 08538,	Word 08579,	Word 08618,	Word 08659,	Word 08698,
	Word 08739,	Word 08778,	Word 08819,	Word 08859,	Word 08900
K0220 DATA	Word 08939,	Word 08980,	Word 09019,	Word 09060,	Word 09101,
	Word 09141,	Word 09180,	Word 09221,	Word 09262,	Word 09301
K0230 DATA	Word 09343,	Word 09382,	Word 09423,	Word 09464,	Word 09503,
	Word 09544,	Word 09585,	Word 09625,	Word 09666,	Word 09707
K0240 DATA	Word 09746,	Word 09788,	Word 09827,	Word 09868,	Word 09909,
	Word 09949,	Word 09990,	Word 10031,	Word 10071,	Word 10112
K0250 DATA	Word 10153,	Word 10194,	Word 10234,	Word 10275,	Word 10316,
	Word 10356,	Word 10397,	Word 10439,	Word 10480,	Word 10519
K0260 DATA	Word 10560,	Word 10602,	Word 10643,	Word 10683,	Word 10724,
	Word 10766,	Word 10807,	Word 10848,	Word 10888,	Word 10929
K0270 DATA	Word 10971,	Word 11012,	Word 11053,	Word 11093,	Word 11134,
	Word 11176,	Word 11217,	Word 11259,	Word 11300,	Word 11340
K0280 DATA	Word 11381,	Word 11423,	Word 11464,	Word 11506,	Word 11547,
	Word 11587,	Word 11630,	Word 11670,	Word 11711,	Word 11753
K0290 DATA	Word 11794,	Word 11836,	Word 11877,	Word 11919,	Word 11960,
	Word 12001,	Word 12043,	Word 12084,	Word 12126,	Word 12167
K0300 DATA	Word 12208,	Word 12250,	Word 12291,	Word 12333,	Word 12374,
	Word 12416,	Word 12457,	Word 12499,	Word 12539,	Word 12582
K0310 DATA	Word 12624,	Word 12664,	Word 12707,	Word 12747,	Word 12789,
	Word 12830,	Word 12872,	Word 12914,	Word 12955,	Word 12997
K0320 DATA	Word 13039,	Word 13060,	Word 13122,	Word 13164,	Word 13205,
	Word 13247,	Word 13289,	Word 13330,	Word 13372,	Word 13414
K0330 DATA	Word 13457,	Word 13497,	Word 13539,	Word 13582,	Word 13624,
	Word 13664,	Word 13707,	Word 13749,	Word 13791,	Word 13833
K0340 DATA	Word 13874,	Word 13916,	Word 13958,	Word 14000,	Word 14041,
	Word 14083,	Word 14125,	Word 14166,	Word 14208,	Word 14250
K0350 DATA	Word 14292,	Word 14335,	Word 14377,	Word 14419,	Word 14461,
	Word 14503,	Word 14545,	Word 14586,	Word 14628,	Word 14670
K0360 DATA	Word 14712,	Word 14755,	Word 14797,	Word 14839,	Word 14881,
	Word 14923,	Word 14964,	Word 15006,	Word 15048,	Word 15090
K0370 DATA	Word 15132,	Word 15175,	Word 15217,	Word 15259,	Word 15301,
	Word 15343,	Word 15384,	Word 15426,	Word 15468,	Word 15510

| 'tC | +0 | +1 | +2 | +3 | +4 |
	+5	+6	+7	+8	+9
K0380 DATA	Word 15554,	Word 15596,	Word 15637,	Word 15679,	Word 15721,
	Word 15763,	Word 15805,	Word 15849,	Word 15891,	Word 15932
K0390 DATA	Word 15974,	Word 16016,	Word 16059,	Word 16102,	Word 16143,
	Word 16185,	Word 16228,	Word 16269,	Word 16312,	Word 16355
K0400 DATA	Word 16396,	Word 16439,	Word 16481,	Word 16524,	Word 16565,
	Word 16608,	Word 16650,	Word 16693,	Word 16734,	Word 16777
K0410 DATA	Word 16820,	Word 16861,	Word 16903,	Word 16946,	Word 16989,
	Word 17030,	Word 17074,	Word 17115,	Word 17158,	Word 17201
K0420 DATA	Word 17242,	Word 17285,	Word 17327,	Word 17370,	Word 17413,
	Word 17454,	Word 17496,	Word 17539,	Word 17582,	Word 17623
K0430 DATA	Word 17667,	Word 17708,	Word 17751,	Word 17794,	Word 17836,
	Word 17879,	Word 17920,	Word 17963,	Word 18006,	Word 18048
K0440 DATA	Word 18091,	Word 18134,	Word 18176,	Word 18217,	Word 18260,
	Word 18303,	Word 18346,	Word 18388,	Word 18431,	Word 18472
K0450 DATA	Word 18515,	Word 18557,	Word 18600,	Word 18643,	Word 18686,
	Word 18728,	Word 18771,	Word 18812,	Word 18856,	Word 18897
K0460 DATA	Word 18940,	Word 18983,	Word 19025,	Word 19068,	Word 19111,
	Word 19153,	Word 19196,	Word 19239,	Word 19280,	Word 19324
K0470 DATA	Word 19365,	Word 19408,	Word 19451,	Word 19493,	Word 19536,
	Word 19579,	Word 19621,	Word 19664,	Word 19707,	Word 19750
K0480 DATA	Word 19792,	Word 19835,	Word 19876,	Word 19920,	Word 19961,
	Word 20004,	Word 20047,	Word 20089,	Word 20132,	Word 20175
K0490 DATA	Word 20218,	Word 20260,	Word 20303,	Word 20346,	Word 20388,
	Word 20431,	Word 20474,	Word 20515,	Word 20559,	Word 20602
K0500 DATA	Word 20643,	Word 20687,	Word 20730,	Word 20771,	Word 20815,
	Word 20856,	Word 20899,	Word 20943,	Word 20984,	Word 21027
K0510 DATA	Word 21071,	Word 21112,	Word 21155,	Word 21199,	Word 21240,
	Word 21283,	Word 21326,	Word 21368,	Word 21411,	Word 21454
K0520 DATA	Word 21497,	Word 21540,	Word 21582,	Word 21625,	Word 21668,
	Word 21710,	Word 21753,	Word 21795,	Word 21838,	Word 21881
K0530 DATA	Word 21923,	Word 21966,	Word 22009,	Word 22051,	Word 22094,
	Word 22137,	Word 22178,	Word 22222,	Word 22265,	Word 22306
K0540 DATA	Word 22350,	Word 22393,	Word 22434,	Word 22478,	Word 22521,
	Word 22562,	Word 22606,	Word 22649,	Word 22690,	Word 22734
K0550 DATA	Word 22775,	Word 22818,	Word 22861,	Word 22903,	Word 22946,
	Word 22989,	Word 23032,	Word 23074,	Word 23117,	Word 23160
K0560 DATA	Word 23202,	Word 23245,	Word 23288,	Word 23330,	Word 23373,
	Word 23416,	Word 23457,	Word 23501,	Word 23544,	Word 23585
K0570 DATA	Word 23629,	Word 23670,	Word 23713,	Word 23757,	Word 23798,
	Word 23841,	Word 23884,	Word 23926,	Word 23969,	Word 24012
K0580 DATA	Word 24054,	Word 24097,	Word 24140,	Word 24181,	Word 24225,
	Word 24266,	Word 24309,	Word 24353,	Word 24394,	Word 24437
K0590 DATA	Word 24480,	Word 24523,	Word 24565,	Word 24608,	Word 24650,
	Word 24693,	Word 24735,	Word 24777,	Word 24820,	Word 24863
K0600 DATA	Word 24905,	Word 24948,	Word 24990,	Word 25033,	Word 25075,
	Word 25118,	Word 25160,	Word 25203,	Word 25245,	Word 25288

Lookup Tables

'tC	+0 +5	+1 +6	+2 +7	+3 +8	+4 +9
K0610 DATA	Word 25329, Word 25542,	Word 25373, Word 25585,	Word 25414, Word 25626,	Word 25457, Word 25670,	Word 25500, Word 25711
K0620 DATA	Word 25755, Word 25967,	Word 25797, Word 26009,	Word 25840, Word 26052,	Word 25882, Word 26094,	Word 25924, Word 26136
K0630 DATA	Word 26178, Word 26390,	Word 26221, Word 26432,	Word 26263, Word 26475,	Word 26306, Word 26516,	Word 26347, Word 26559
K0640 DATA	Word 26602, Word 26814,	Word 26643, Word 26856,	Word 26687, Word 26897,	Word 26728, Word 26940,	Word 26771, Word 26983
K0650 DATA	Word 27024, Word 27236,	Word 27067, Word 27277,	Word 27109, Word 27320,	Word 27152, Word 27362,	Word 27193, Word 27405
K0660 DATA	Word 27447, Word 27658,	Word 27489, Word 27700,	Word 27531, Word 27742,	Word 27574, Word 27784,	Word 27616, Word 27826
K0670 DATA	Word 27868, Word 28079,	Word 27911, Word 28120,	Word 27952, Word 28163,	Word 27995, Word 28204,	Word 28036, Word 28246
K0680 DATA	Word 28289, Word 28457,	Word 28332, Word 28500,	Word 28373, Word 28583,	Word 28416, Word 28626,	Word 28416, Word 28667
K0690 DATA	Word 28710, Word 28919,	Word 28752, Word 28961,	Word 28794, Word 29003,	Word 28835, Word 29045,	Word 28877, Word 29087
K0700 DATA	Word 29129, Word 29338,	Word 29170, Word 29379,	Word 29213, Word 29422,	Word 29254, Word 29463,	Word 29297, Word 29506
K0710 DATA	Word 29548, Word 29757,	Word 29589, Word 29798,	Word 29631, Word 29840,	Word 29673, Word 29882,	Word 29715, Word 29923
K0720 DATA	Word 29964, Word 30173,	Word 30007, Word 30214,	Word 30048, Word 30257,	Word 30089, Word 30298,	Word 30132, Word 30341
K0730 DATA	Word 30382, Word 30589,	Word 30423, Word 30632,	Word 30466, Word 30673,	Word 30507, Word 30714,	Word 30548, Word 30757
K0740 DATA	Word 30797, Word 31006,	Word 30839, Word 31047,	Word 30881, Word 31088,	Word 30922, Word 31129,	Word 30963, Word 31172
K0750 DATA	Word 31213, Word 31420,	Word 31254, Word 31461,	Word 31295, Word 31504,	Word 31338, Word 31545,	Word 31379, Word 31585
K0760 DATA	Word 31628, Word 31833,	Word 31669, Word 31876,	Word 31710, Word 31917,	Word 31751, Word 31957,	Word 31792, Word 32000
K0770 DATA	Word 32040, Word 32246,	Word 32082, Word 32289,	Word 32124, Word 32329,	Word 32164, Word 32371,	Word 32206, Word 32411
K0780 DATA	Word 32453, Word 32659,	Word 32495, Word 32700,	Word 32536, Word 32742,	Word 32577, Word 32783,	Word 32618, Word 32824
K0790 DATA	Word 32865, Word 33070,	Word 32905, Word 33110,	Word 32947, Word 33152,	Word 32987, Word 33192,	Word 33029, Word 33234
K0800 DATA	Word 33274, Word 33479,	Word 33316, Word 33521,	Word 33356, Word 33561,	Word 33398, Word 33603,	Word 33439, Word 33643
K0810 DATA	Word 33685, Word 33889,	Word 33725, Word 33929,	Word 33767, Word 33970,	Word 33807, Word 34012,	Word 33847, Word 34052
K0820 DATA	Word 34093, Word 34296,	Word 34134, Word 34338,	Word 34174, Word 34378,	Word 34216, Word 34420,	Word 34256, Word 34460
K0830 DATA	Word 34500, Word 34704,	Word 34542, Word 34744,	Word 34582, Word 34786,	Word 34622, Word 34826,	Word 34664, Word 34866

	+0 +5	+1 +6	+2 +7	+3 +8	+4 +9
K0840 DATA	Word 34908, Word 35109,	Word 34948, Word 35151,	Word 34999, Word 35192,	Word 35029, Word 35231,	Word 35070, Word 35273
K0850 DATA	Word 35313, Word 35515,	Word 35353, Word 35555,	Word 35393, Word 35595,	Word 35435, Word 35637,	Word 35475, Word 35676
K0860 DATA	Word 35718, Word 35920,	Word 35758, Word 35960,	Word 35798, Word 36000,	Word 35839, Word 36041,	Word 35879, Word 36081
K0870 DATA	Word 36121, Word 36323,	Word 36162, Word 36363,	Word 36202, Word 36403,	Word 36242, Word 36443,	Word 36282, Word 36484
K0880 DATA	Word 36524, Word 36725,	Word 36564, Word 36765,	Word 36603, Word 36804,	Word 36643, Word 36844,	Word 36685, Word 36886
K0890 DATA	Word 36924, Word 37125,	Word 36965, Word 37165,	Word 37006, Word 37206,	Word 37045, Word 37246,	Word 37085, Word 37286
K0900 DATA	Word 37326, Word 37526,	Word 37366, Word 37566,	Word 37406, Word 37606,	Word 37446, Word 37646,	Word 37486, Word 37686
K0910 DATA	Word 37725, Word 37925,	Word 37765, Word 37965,	Word 37805, Word 38005,	Word 37845, Word 38044,	Word 37885, Word 38084
K0920 DATA	Word 38124, Word 38323,	Word 38164, Word 38363,	Word 38204, Word 38402,	Word 38243, Word 38442,	Word 38283, Word 38482
K0930 DATA	Word 38521, Word 38719,	Word 38561, Word 38759,	Word 38600, Word 38798,	Word 38640, Word 38838,	Word 38679, Word 38878
K0940 DATA	Word 38917, Word 39115,	Word 38957, Word 39164,	Word 38996, Word 39195,	Word 39036, Word 39234,	Word 39076, Word 39274
K0950 DATA	Word 39314, Word 39511,	Word 39353, Word 39549,	Word 39393, Word 39590,	Word 39432, Word 39628,	Word 39470, Word 39668
K0960 DATA	Word 39707, Word 39905,	Word 39746, Word 39944,	Word 39786, Word 39984,	Word 39826, Word 40023,	Word 39865, Word 40061
K0970 DATA	Word 40100, Word 40298,	Word 40140, Word 40337,	Word 40179, Word 40375,	Word 40219, Word 40414,	Word 40259, Word 40454
K0980 DATA	Word 40493, Word 40689,	Word 40533, Word 40728,	Word 40572, Word 40765,	Word 40610, Word 40807,	Word 40651, Word 40846
K0990 DATA	Word 40885, Word 41081,	Word 40924, Word 41119,	Word 40963, Word 41158,	Word 41002, Word 41198,	Word 41042, Word 41237
K1000 DATA	Word 41276, Word 41470,	Word 41315, Word 41509,	Word 41354, Word 41548,	Word 41393, Word 41587,	Word 41431, Word 41626
K1010 DATA	Word 41665, Word 41859,	Word 41704, Word 41898,	Word 41743, Word 41937,	Word 41781, Word 41976,	Word 41820, Word 42014
K1020 DATA	Word 42053,	Word 42092,	Word 42131,	Word 42169	

Chapter 13

Educating with Model Rocketry

Project 48: Educating with Model Rocketry

You will need

- Large quantity of children
- Large quantity of rocket kits

Tools

- Blackboard
- Chalk

Model rocketry is a microcosm of what happens in "real world" aerospace activities, and as such it can be used as a tool for delivering education with real meaning and a direct bearing on the real world.

The prospects of building and launching model rockets is enough to motivate even the most reluctant student to become enthusiastic about the subject. This motivation can be channeled to make them learn about the "satellite subjects" first before concentrating on the construction of their rockets.

The trick to educating with model rocketry is get students to learn stuff before they launch the rockets. Build and launch the rockets first and try and follow up with some wholesome education and you won't stand a chance. If you start with all the "heavy" stuff first of all, the rocketry acts as a reward for all their hard work and endeavours.

Below are some subject-specific ideas for model rocket education

Business studies

Students can contrast the state-funded space programs of different countries and nations with the commercial aerospace industry. Students can look at the demand for space travel and the companies that supply the products and services to meet that demand. They could explore the future possibilities afforded by space tourism, the economics of space travel and the gains to both consumers and businesses.

Online resources

www.virgingalactic.com/
www.spacefuture.com/tourism/tourism.shtml
www.space.com/spacetourism/

Get students to prepare a "business plan" for a space tourism company; in addition to working out how much to charge for each ticket, get them to "build" a launch vehicle. Give them a "budget" and charge them for all the parts that they use from their imaginary budget. Tell them that each passenger they have to carry weighs the same as, say, a quarter, and that their launch vehicle has got to carry all the passengers to a height of 100 m safely and return them to earth. There is a trade-off between the number of passengers they will put inside their rocket, and how successful the mission is likely to be. The successful team is the one that manages to carry their passengers into "space" and back safely for the least amount of money.

Figure 13-1   *Film-can rockets used in a science lesson.*

Physics and chemistry

Physics can be seen as dull by many students, but there is no reason for this, the subject can be brought alive and made interesting by the inclusion of a little rocket science in lessons. Chapter 2 provides a good springboard for devising a series of experiments that helps illustrate how interesting the subject can really be. Simple experiments like the film-can rockets (Project 2: Figure 13-1) can be used with all ages as they do not entail work with hazardous chemicals – everything being commonly available in most households.

English

Space and rocketry is a great launchpad for exploring English. There are lots of great works of English literature that focus on space, rocketry, and other worlds. Students can read some of the great works such as *War of the Worlds* by H.G. Wells, and explore the relationship between the written word and space exploration. There is also a lot of scope for creative writing.

You could call a local paper, and have a competition between all the students for the best written launch report – the best gets published in the local rag. Even if the articles don't hit the big time, there is still plenty of scope for promotion in school newspapers and publications. It never hurts to raise the banner of the school and increase the school's profile.

Design and technology

Product design

Students can explore designing and marketing an innovative new model rocket kit. There is a wealth of information on existing products that can be researched by students; the design and construction of the kit poses challenges for both materials and manufacturing methods – how can the product be produced in batches, what manufacturing methods will be used to make large quantities of nose cones, fin transitions, motor hooks, etc.

Graphic communication

There is plenty of scope for using model rocketry as an activity for graphics lessons; designing packaging for a new model rocket kit is a good project to engage students; designing clear instructions for the assembly of a kit is another. Exploded views of model rockets form the foundation of an interesting challenging technical drawing project.

The wide variety of materials used in model rocketry provide-scope for numerous hand rendering tasks. Computer-aided design (CAD) programs can be used by students to prepare 3D computer renderings of their models. One popular piece of software for use in schools is Pro/DESKTOP, which provides an affordable CAD solution, and is used by Airbus, Hughes Space & Communications, Boeing, NASA, Ratheon, and BAE Systems. For educational use, the software is highly affordable.

Online resources

Pro Desktop is available from
http://www.ptc.com/

Citizenship education

Space exploration has a lot of moral, ethical, and social concerns connected with it – is it ethical to explore strange new worlds, and how could we affect alien cultures if we landed on another planet? Should humans be sent into space? Is it reasonable to ask someone to risk their life in the name of human endeavour? What are the environmental impacts of the space industry? Is the pollution caused justified by the outcomes?

Young rocketeers can learn about responsible rocketry and their obligations towards others in the wider community when launching their rockets.

Media studies

There is a lot of scope for using rocketry in mixed media projects. Students can look at rocketry and the media; how is rocketry portrayed and has that changed in the 21st century compared to say the space race and the cold war? Students could make a short film about the space race, model rocketry, or another space-related topic of their choosing.

History

The history of space exploration is an interesting one. Students can combine learning about significant events from the history of rocketry with building scale models of the rockets that changed the world. The research required to build accurate scale models will all feed into their knowledge about space exploration in times gone by.

Tips and tricks for rocket construction in classrooms

Because of time constraints, it is often the case that rockets need to be built and launched within the same afternoon. To save waiting about for adhesives to dry fully, a tube of cyanoacrylate adhesive and a can of spray activator speed up the process considerably. Get students to tack the fins in place with white glue, and then when the glue has set a little, a couple of dabs on each side of each fin will ensure that the fins are mounted securely enough to fly the same day. By using an activator, the cyanoacrylate adhesive is set for use within minutes.

Because model rockets invariably cost money, getting students to work in pairs on the rocket invariably brings costs down significantly (Figure 13-2).

Figure 13-2 *Students hard at work constructing their model rockets.*

Figure 13-3 *Students share the task of launching and tracking their rocket.*

Without fail there will be some squabbles about who "presses the button"; by giving one student a tracking device and the other the responsibility of launching the rocket, the tasks are shared and there is less likely to be a problem. This keeps both students occupied and no-one feels hard-done-by. The next time the rockets are launched, roles can be exchanged.

The day of launch is consistently an exciting time for all concerned. For this reason, it is important that everyone keeps calm and the students are well aware of the rules and their responsibilities.

A large school field provides a suitable launch site; sports pitches are especially good as there is

Figure 13-4   *Students prepare rockets for launch.*

Figure 13-5   *The button is pressed – students watch in amazement.*

Figure 13-6   *Recovering rockets – see how they run!*

generally a white line that everyone can be told to stand behind.

For smaller school fields, you will need to find which direction the wind is blowing in and "weathercock" the model so that it can be recovered within the premises. Launching projectiles at neighbouring houses is a quick way of lowering the school's reputation in the community!

Depending on the age of the students, it may be felt appropriate for the tutors to load the rocket engines and the igniters. Regardless, keep these under lock and key until the time of launch to prevent any errant horseplay, which could prove highly dangerous.

Once all of the students are ready, get them all to be quiet while the group currently launching gives a loud audible countdown. The moment the button is pressed for the first time, the students will truly realize that the fruits of their labor are worthwhile (Figure 13-5).

Unless you have a lot of time on your hands, save recovery until the end. Ensure that all students remember where their rockets land, and remind students, in their anxiety to recover their own models, not to tread on anyone else's as they run across the field.

Chapter 14

Model Rocketry Clubs

Project 49: Starting a Model Rocketry Club

Starting a model rocketry club is a satisfying way of sharing your hobby with a large group of people, meeting socially for gatherings, and displaying your hobby to members of the general public.

You can start a club easily or you can start it properly. Starting a club properly means that you do things correctly from the beginning and don't leave anything to chance! Make sure you get permission to use the launch site you are planning to fly from, and obtain some public liability insurance. A cost effective way of doing this is by becoming a "chartered club" of one of the national organisations, such as the National Association of Rocketry the (NAR).

Try and get involved in the community as much as possible and promote the hobby – this way you will rapidly gain new members. Below are some ideas on how to get involved.

Figure 14-1 *Rocketeers getting involved with local schools. Image courtesy Peter Barratt.*

them in the hobby. See Figure 14-1 for a plethora of rockets made by schoolchildren in co-operation with a local rocket club.

Getting featured in the media

Local newspapers

Local newspapers are a good place to get promoted – they are always looking for interesting material to fill their pages, so an event such as getting involved with a local school is likely to attract positive attention to the club. Get involved with your community or run a workshop day at a local school – quite often teachers will be receptive to bringing in an outside "expert" to show their class something new and engage

Local radio

Local radio is another way that you can get your rocket club publicised – offer to do an interview with a local DJ on the hobby and what is involved. Throw in a few tracks from my top-10 playlist and you are guaranteed a great afternoon of space entertainment!

Top-10 space playlist

Rocket Man Elton John

Space Oddity David Bowie

Space Truckin' Deep Purple

Calling Occupants of Interplanetary Craft The Carpenters

Miles High Eight The Byrds

Walking on the Moon The Police

Star Trekking The Firm

Ziggy Stardust David Bowie

Walking on the Milky Way OMD

Above the Clouds ELO

Figure 14-2 *Local radio station LINK FM doing a report on the HART rocket team. Image courtesy Peter Barratt.*

Television

If you get the chance, television is a great way to demonstrate your rocket club's skills to the world at large. Rocketry is a quirky hobby, one that not everyone knows about. If you are holding a big event, make sure you make a cheeky phone call to the local TV news – don't be shy, news crews often run short features on local interest. Go forth and promote the hobby! You might even be a star!

Another angle that you might like to consider is that there are frequently advertisements for programs that require entrants with a certain "technical bent." Figure 14-4 illustrates the sort of coverage that can be gotten by a rocket club. Here Hornchurch Airfield Rocket Team (HART) enters the "Eggspress" in the *Techno Games* TV tournament.

Charity events

Rocketry is a great launch vehicle for many charity events; with a little teamwork, your rocket crew can successfully raise a stack of cash for good causes. Consider ways that your rocket team can get involved in the community. Figure 14-5 shows the Red Nose Day Rocket built by HART to raise money for charity.

Getting the family involved

The chances are that whether your family find space, science, and technology interesting or not, they will relish the chance of joining in at a club launch day.

Figure 14-3 *Blue Sky TV filming. Image courtesy Peter Barratt.*

Even if your family aren't remotely interested in rockets, the chances are they will find their own niche, whether that be baking cakes, making tea or coffee, or tending the barbecue! Everyone has something to bring to the party, and it is important that you remember to thank the "honorary" members of the club, who, though they may not build or launch the rockets, work behind the scenes to keep the club moving and create a pleasant atmosphere for all.

Passing on the baton

Rocketeers, whether old or young, big or small, from those using $\frac{1}{3}$ As to Ns all have something to learn from

Figure 14-4 *HART rocket featured on Techno Games TV tournament. Image courtesy Peter Barratt.*

Figure 14-7 *Rockets big and small coexist happily on the launch pad. Image courtesy Peter Barratt.*

Figure 14-5 *The Red Nose Day rocket 1999. Image courtesy Peter Barratt.*

Figure 14-8 *Multiple launches are a great spectacle. Image courtesy Gary White.*

Figure 14-6 *Rocketry is a family activity. Image courtesy Gary White.*

each other. Model rocketry clubs are a great way of sharing knowledge, working on problems collaboratively and passing on the baton from one generation to the next. Whether your rocket is big or small, it will coexist happily on the club launchpad with all the other rockets (See Figure 14-7).

Pulling the crowds

There is nothing like a crowd pleaser! Attracting new members to your club is imperative for its continued survival, it is important to make your club visible to the local community in order to attract new members.

Scout groups

Model rocketry is a very worthwhile activity for scouts as it gives them the opportunity to work towards the Space Exploration Merit Badge.

Project 50: Space Exploration Merit Badge

In order to earn this badge, a scout must demonstrate that s/he has knowledge of the historical development of space exploration, the goals that can be achieved by exploring space, and the knowledge gained by this endeavour. Furthermore, the scout must demonstrate that they have knowledge of the benefit to mankind of space exploration.

The scout must then design a collector's card of their favourite space pioneer. Professor Colin Pillinger would be my first choice!

The scout must then build and launch a model rocket demonstrating knowledge of the following components:

- body tube;
- engine mount;
- fins;
- igniter;
- launch lug;
- nose cone;
- payload;
- recovery system;
- rocket engine.

The scout must then demonstrate knowledge of the following rocket science:

- the law of forces – action–reaction;
- how rocket engines work;

Figure 14-9 *Space Exploration Merit Badge.*

- how satellites stay in orbit;
- how satellite pictures of the Earth and pictures of other planets are made and transmitted.

The scout must then pick two out of the following activities:

- compare and contrast an early manned and unmanned space mission;
- make a scrapbook of a current space mission;
- design an unmanned space mission.

The scout must then describe the importance of either the Space Shuttle or the International Space Station.

The scout must then design a "Moon base" or "Mars base" looking specifically at:

- energy source;
- construction;
- life support;
- purpose.

The scout must then highlight two careers in connection with the space industry and highlight the qualifications required and the progression route.

So buy this book for a scout you love!

Online resources

Some good online resources for scouts wishing to earn this badge:

my.execpc.com/~culp/space/space.html
www.meritbadge.com/mb/107.htm
jrm.phys.ksu.edu/Scouts/
jrm.phys.ksu.edu/Scouts/sld001.htm
forums.seds.org/showthread.php?t=439
illinois.seds.org/edu/Scouts_in_Space.pdf

Model Rocket Safety

Launching rockets is one thing; launching rockets safely is quite another. Safety is absolutely paramount, and has been stressed throughout this book. Launching rockets is safe – if done correctly. In the hands of fools, model rockets can be very dangerous. Anyone can launch a rocket safely; all that is needed is a little forethought, and the time taken to follow procedure and do things correctly.

Reprinted below are the safety codes of the National Association of Rocketry (NAR), the largest rocketry club in the US, and the United Kingdom Rocketry Association (UKRA), the largest rocketry association in the UK.

Whether you are new to the hobby, or an experienced rocketeer, joining clubs has a lot to offer you. They organize properly run, safe events; often as a member you will qualify for such benefits as third-party insurance.

National Association of Rocketry (US) Model Rocket Safety Code

1. Materials. I will use only lightweight, non-metal parts for the nose, body, and fins of my rocket.

2. Motors. I will use only certified, commercially-made model rocket motors, and will not tamper with these motors or use them for any purposes except those recommended by the manufacturer.

3. Ignition System. I will launch my rockets with an electrical launch system and electrical motor igniters. My launch system will have a safety interlock in series with the launch switch, and will use a launch switch that returns to the "off" position when released.

4. Misfires. If my rocket does not launch when I press the button of my electrical launch system, I will remove the launcher's safety interlock or disconnect its battery, and will wait 60 seconds after the last launch attempt before allowing anyone to approach the rocket.

5. Launch Safety. I will use a countdown before launch, and will ensure that everyone is paying attention and is a safe distance of at least 15 feet away when I launch rockets with D motors or smaller, and 30 feet when I launch larger rockets. If I am uncertain about the safety or stability of an untested rocket, I will check the stability before flight and will fly it only after warning spectators and clearing them away to a safe distance.

6. Launcher. I will launch my rocket from a launch rod, tower, or rail that is pointed to within 30 degrees of the vertical to ensure that the rocket flies nearly straight up, and I will use a blast deflector to prevent the motor's exhaust from hitting the ground. To prevent accidental eye injury, I will place launchers so that the end of the launch rod is above eye level or will cap the end of the rod when it is not in use.

7. Size. My model rocket will not weigh more than 1500 grams (53 ounces) at liftoff and will not contain more than 125 grams (4.4 ounces) of propellant or 320 N-sec (71.9 pound-seconds) of total impulse. If my model rocket weighs more than one pound (453 grams) at liftoff or has more than four ounces (113 grams) of propellant, I will check and comply with Federal Aviation Administration regulations before flying.

8. Flight Safety. I will not launch my rocket at targets, into clouds, or near airplanes, and will not put any flammable or explosive payload in my rocket.

9. Launch Site. I will launch my rocket outdoors, in an open area at least as large as shown in the accompanying table, and in safe weather conditions with wind speeds no greater than 20 miles per hour. I will ensure that there is no dry grass close to the launch pad, and that the launch site does not present risk of grass fires.

10. Recovery System. I will use a recovery system such as a streamer or parachute in my rocket so that it returns safely and undamaged and can be flown again, and I will use only flame-resistant or fireproof recovery system wadding in my rocket.

11. Recovery Safety. I will not attempt to recover my rocket from power lines, tall trees, or other dangerous places.

Table A-1

Launch site dimensions

Installed total impulse (N-sec)	Equivalent motor type	Minimum site dimensions
0.00–1.25	1/4A, 1/2A	50
1.26–2.50	A	100
2.51–5.00	B	200
5.01–10.00	C	400
10.01–20.00	D	500
20.01–40.00	E	1000
40.01–80.00	F	1000
80.01–160.00	G	1000
160.01–320.00	Two Gs	1500

The latest version is available online at:

www.nar.org/NARmrsc.html

The version printed here is the February 2001 revision.

United Kingdom Rocketry Association Safety Code

This is the Short Version of the UKRA Safety Code; the full version should be downloaded and printed for reference, and where there is any discrepancy between the full and short codes, the full code should be adhered to.

1 General rules

1.1 Safety

Safety is the concern of all members. Members causing serious damage/injury to third parties, livestock, vehicles or property whilst involved in Rocketry of any kind must report the incident in full to the Safety Committee, even if the UKRA codes of practice were not in force at the time of the incident.

1.2 Payloads

No UKRA member's rocket will ever carry live animals or any payload that is intended to be flammable, explosive, or harmful.

2 Equipment

2.1 The Rocket

All rockets flown under this safety code shall be made of lightweight materials such as paper, wood, rubber or plastic. The range safety officer must be satisfied that the rocket is flight worthy and sufficiently robust to survive launch, aerodynamic, and recovery system loads. All rockets must use a recovery system that will return it safely to the ground so it may be flown again.

2.2 Motors

The motor must only be used in the manner recommended by the manufacturer. It is not permissible to alter the rocket motor, its parts or its ingredients in anyway.

2.3 Igniters

The Safety Officer must be satisfied with the igniter system that is connected to the rocket motor.

2.4 Weight & Power

Any rocket must not have a mass greater than the manufacturer's recommended maximum lift-off mass for the motors used.

2.5 Launch Controller

The launch controller must include a safety key to immobilise the system when removed. This key should only be in place at the time of the launch and is to be removed immediately after an ignition attempt, especially in the event of a misfire.

3 The Launch Site

3.1 Safety Officer

The Safety Officer has authority over and above all other persons present at the Launch Site and has the power to delay or cancel any launch until satisfied that it can proceed safely.

3.2 Personnel

Only UKRA members may approach nearer the rocket than the minimum Safe Distance during or after an igniter is being/has been installed into the Rocket Motor(s). Members may only approach nearer than the Safe Distance with the approval of the Safety Officer.

3.2.1 Spectators

All spectators/onlookers/press at a UKRA launch must be kept at least the minimum Safe Distance away from the launch area as determined by the total impulse of the Rocket Motor(s) according to the Safe Distance Table.

3.2.2 Minders

Any persons at the launch site who cannot watch the rocket, e.g. due to their monitoring of equipment must be protected; either by a physical safety barrier or by persons beside them who can watch the rocket and issue a warning or take protective action.

3.2.3 Visual Rule

All persons at the launch site should be aware that for their own safety they must keep their eyes on the rocket from at least two seconds before launch until either the rocket lands or until visual contact is lost.

3.2.4 Safe Distance Table

All persons, except those required for the launch of a rocket should kept at least the given minimum distance from the Rocket Motor during/after igniter installation.

4 Flying

4.1 Launch Permission

Before launching, a UKRA member must obtain the permission to launch from the range Safety Officer.

4.2 Launching

A clearly audible countdown of at least five seconds must be given, either by the launch person, the Safety Officer or any person recognised by all present as responsible for the countdown and authorised by the Safety Officer.

4.3 Misfires

If a Rocket suffers a misfire, no one may approach the Launch Pad until waiting for one minute, after this time the Safety Officer should give permission for one person to approach the Rocket.

The latest FULL version is available online from:

www.ukra.org.uk/docs/Safetycode_v421.pdf.

International Listing of Model Rocket Clubs

I have included links to the following clubs' websites, with addresses only for the major clubs in any one country. Listing all the addresses would quickly fill the book up and wouldn't represent good value for money!

North America

United States of America

National Association of Rocketry

http://www.nar.org/

Canada

Association Astronautique Amateur du Québec.

http://www.game-master.com/yves/index.html

British Columbia Rocketry Club http://bcrc.ca/

Ottawa Rocketry Group http://www.ottawa-rocketry.org/

Europe

Albania

Albanian Eagles (http://www.albanian-eagles.org)

Belgium

Vlaamse Raket Organisatie http://www.vro.be/

Denmark

Danish Amateur Rocketry Club http://www.dark.dk/

Germany

Aero Club Rheidt (Germany) (http://www.ac-r.de)

Interessengemeinschaft Modellraketen e.V. (IMR) (http://www.modellraketen.org)

Muenchner Modellraketenverein e.V. (MMV) (http://www.modellraketen.de/mmv)

Netherlands

Dutch Rocket Research Association http://www.drra.nl/

Navro http://www.navro.nl/

Slovenia

Astronavstko Raketarski Klub Vladimir M. Komarov (ARK Komarov)

www.komarov.vesolje.net

Spain

Aeromodelling club Tamaran

(http://www.terra.es/personal8/albparra)

Africa

South Africa

South African Rocketry http://www.sarocketry.co.za/

South African Amateur Space Administration

http://www.sahpr.org.za/

Asia

Japan

Japan Association of Rocketry

http://www.ja-r.net/

Oceania

Australia

Hi Flyte Club Inc http://users.chariot.net.au/~hiflyte/

New South Wales Rocketry Association

http://www.users.bigpond.com/pagilchrist/main/index1.html

Queensland Advanced Rocketry Club

http://www.varietyqld.org/QARC/index_main.htm

New Zealand

Hutt Valley Rocket Club

http://www.geocities.com/nzrocketman/

New Zealand Rocketry

http://www.rockets.co.nz/

3FNC rocketeer slang for a dead easy rocket. (Three Fins and a Nose Cone!)

apogee the point in a Rocket's flight when it is at its highest altitude.

AWG American wire gauge – this number specifies how thick your wire is.

blast deflector the metal plate on a rocket launch pad that prevents the hot exhaust gases from burning the rest of the launch pad.

CATO catastrophic take off. When everything goes. wrong and your rocket blows up on the launch pad. To paraphrase the great Peter Sellers "Oh no... not now CATO".

CG centre of gravity.

clustered engines where more than one rocket engine is used at the same time to produce greater thrust than could be accomplished with a single engine.

CNC computer numerical control, made by computer from PC drawings.

CP centre of pressure.

deceleration rate of decline in speed of an object.

de Laval nozzle the shape that produces maximum thrust from the combustion products of a rocket motor, see Chapter 2: Rocket Science.

delay charge the piece of propellant in a model rocket motor that burns slowly, allowing the motor to coast.

ejection the point when the small explosive charge in the rocket engine creates a large volume of gas inside the rocket. This causes it to pressurize and thereby deploys the recovery system.

igniter the electrical fuse that is used to safely ignite a rocket motor.

launch lug a small tube (or "pip" in the case of rail launchers) that guides the rocket up the launch device during its first few seconds of flight.

launch rail the rail that guides the model rocket during its first few seconds of flight until it has gained sufficient speed that its fins can keep it stable. Generally launch rails are used for the additional strength to carry larger rockets. A pip is used to guide the rocket up the rail.

launch rod similar to a launch rail, but is generally used where less strength is required, such as with smaller model rockets. A small tube or piece of a straw is affixed to the rocket to form a launch lug.

LES Lilliput Edison screw. A type of thread on small indicator bulbs.

pitch the left or right movement of a rocket's nose about its vertical axis.

single-station tracking the simplest tracking system, where a single station is employed.

staging using a "series" of rockets bonded together, in either a parallel or series configuration, which provides more boost at the start of a rocket's flight with the booster stages being discarded by the rocket as the flight progresses.

streamer a recovery device for smaller rockets.

three-station tracking a tracking system where three trackers are used for greater accuracy than a two-station system.

throat the middle section of the de Laval nozzle which is the narrowest, and where the most restriction occurs.

turbulent flow uneven air movement over the model rocket's "skin". Eddy currents form around uneven surfaces.

two-station tracking a tracking system where two trackers are used for greater accuracy than a single-station system.

velocity distance moved per unit time ("speed").

weathercocking a process whereby the rocket is turned to face into the wind away from a vertical path.

wind tunnel a pipe through which air can be channelled and controlled to produce a laminar flow at a constant velocity, allowing a rocket's aerodynamics to be tested under repeatable conditions.

wood grain the direction of the fibres in a piece of wood. Important for rocketeers to note in order to build strong components from wood. The strength of the wood is perpendicular to the direction of the fibres.

yaw a backwards and forwards motion of the rocket where the direction is seen as left to right. *See also* pitch.

zippering a situation that arises when a rocket's recovery system deploys at the wrong time, i.e. when the airframe is moving at high speed. The drag caused by the recovery device causes the shock cord to become tight and cut through the model like a wire through cheese.

Index

Index

construction, 41–53
education with, see Education and
    model rocketry
fins, see Fins
flight computers, see Flight computers
igniters, see Igniters
launching, see Launch techniques
materials, see Materials
nose cones, see Nose cones
payload-carrying, see Payload(s)
recovery systems, see Recovery systems
science, 9–20
spacing rings, see Spacing rings
stability, see Stability
staged, see Staged rockets
transitions, see Transitions
tools, see Tools
UFOs, see UFOs
workshop, 21–33
Moon landing, 6
Motion, Laws of, 14, 90, 178

N

Newton, Isaac, 14
    Laws of Motion, 14
Night tracer rocket
    construction of, 60–2
    EL inverter, 62
    electroluminescent panels, 61–2
Nose cones, 19, 41–2, 178
    blunt, 19
    cone, 19
    drag, and, 19
    hemisphere, 19
    length-to-width ratio, 42
    ogive, 19
    parabola, 19
    plastic, 42
    profile, 19

O

Oxygen, production of, 17

P

Paints, see Materials
Painting techniques, 27–8
    balsa, for, 28
    plastic, for, 28
    preparation, 27
Parachutes, see Recovery systems
Parallax
    BS2pe, 121–2

BS2p40, 121–2
    thermocouple kit, 132
Payload(s)
    biological, 6
    camera, see Camera payloads
    payload-carrying rocket, 52–3
        construction of, 52–3
    rocket mail, 117–19
Polymorph, 29–30, 111, 113
Potassium nitrate, for black powder, 10
Propellant quantity, 16
    and distance, 16
Publicity
    charity events, 176
    local newspapers, 175
    local radio, 175
    television, 176

R

Reaction forces, 13–16
    balanced, 13
    chemical, 13
        fast, 15, 16
    equal and opposite, 14
    unbalanced, 13
Reaction mass, 10, 12, 16, 17
Reaction rates, 13–16
    demonstrating, 13–16
Recovery systems, 41, 55–65, 174, 178
    basics, 55
    camera, 108
    drag, and, 20
    ejection charge, 55
    featherweight, 56
    glide, 59
    helicopter (autorotation), 57–9
    night tracer, 60–1
    nose blow, 56
    parachute (parasheet), 47, 53, 57, 58, 108
        diameter ready reckoner, 58
        shroud lines, 57
    recovery wadding, 60
    shock cord, 55, 108
        materials, 60
        mounts, 59
    size, 98–9
    streamer, 56
        rolling techniques, 56–7
    strobe beacon, 63–5
    tumble, 50, 56
Redstone rocket, 5
Rocket mail, 117–19
    history, 118–19
    stamps, 118
Rocket math, 9, 89–99
Rocket motors
    Alka-Seltzer, 14–16

S

T

U

UFOs
  Art Applewhite design, 50–1
  construction of, 50–52
  simple, 51
  featherweight recovery, 56
  tumble recovery, 50, 56

V

V2 rocket, 5
Vibration measurement, 129

W

Wells, H.G, 3, 4
Wings, *see* Fins
Wind tunnel, 30–1, 188
  construction of, 30–1

Z

Zippering, 188

Index

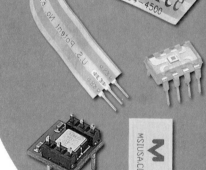

Free My Boid with every starter set purchased with motors from the Model Rocket Shop www.modelrockets.co.uk

Use our web shop to pick the best starter set and motors for your orbital ambitions and mention this FREE MYBOID Coupon Offer in the order comments field and we will send you a free Semroc My Boid high performance rocket kit, FREE of charge. One free My Boid per customer!

Starter sets are a great deal because in addition to an almost Ready To Fly rocket, they have a launch pad, launch controller and full instructions for safe flights … all at a bargain price.

My Boid is a simple, easy to build rocket with a precision turned balsa nose and laser-cut fins. Built with care it will pop up to 350 feet with an A motor. On a C6-5 motor it will soar to over 1000 feet. Outta sight!

My Boids return to earth gently on a parachute and are ready for another blistering ascent with a fresh motor in a few minutes.

Glue for assembly, paint and finishing supplies and launching consumables are not included with the My Boid.

Free SpaceCAD rocket design software CD

Register on www.modelrockets.co.uk and use the web site's contact form to send an email quoting this Coupon Offer. We will send a FREE, unrestricted 30 day copy of SpaceCAD on a CD and a coupon for £5 off your purchase of a full licensed copy.

If you are ready to move on from building kits strictly by the book, SpaceCAD is what you MUST have! SpaceCAD is *the* European rocket design software that is easy enough for almost anyone to get on with but still complex enough to handle many High Power Rocket rocket designs.

Instant Rocket Science!

There is practically NO learning curve! Start with a nose cone, specify the size, shape and material, add a body tube, fins and a motor and press the button for a virtual flight test. SpaceCAD's all menu driven and the rocket science is well hidden beneath an easy to use interface.